Multidisciplinary Approaches for Studying and Combating Microbial Pathogens

Edited by

A. Méndez-Vilas

BrownWalker Press

Boca Raton

Multidisciplinary Approaches for Studying and Combating Microbial Pathogens

BrownWalker Press
Boca Raton, Florida • USA
2015

ISBN-10: 1-62734-544-2
ISBN-13: 978-1-62734-544-6

www.brownwalker.com

Cover image by Eraxion/Bigstock.com

TABLE OF CONTENTS

Introduction

We are pleased to present a selection of papers presented at The III International Conference on Antimicrobial Research (ICAR2014), which was held in Madrid, Spain, from 1 to 3 October 2014.

The aims of this conference were to provide a forum for discussion of cutting edge antimicrobial research and to create an opportunity for microbiologists, biochemists, genetists, clinicians, physicists, engineers... to meet and have the chance to find new research colleagues and partners for future research works. This third edition gathered 444 participants, coming from 58 countries, and nearly 450 works were presented at the conference. Some of those research works are discussed in this book covering the topics: antimicrobial natural products, biofilms, antimicrobial surfaces, antimicrobial resistance and clinical and medical microbiology...

We would like to thank the International Advisory Committee of this conference, a group of international experts who guided us in the development of the conference program. And, of course, we also thank the authors for submitting their work and sharing their findings.

This book serves as formal proceedings of the meeting. We hope readers will find this set of papers inspiring and stimulating in their current research work and look forward to seeing another fruitful edition in 2016.

A. Méndez-Vilas

Editor

ICAR2014 General Coordinator

A clinical resistant isolate of opportunistic fungal pathogen, *Candida albicans* revealed more rigid membrane than its isogenic sensitive isolate

Arvind Kumar[1,€], **V. S. Radhakrishnan**[1,€], **Richa Singh**[1], **Manish Kumar**[1], **Nagendra N. Mishra**[2] and **Tulika Prasad**[1,*]

[1]Advanced Instrumentation Research Facility (AIRF), Jawaharlal Nehru University, New Delhi-110067, India.
[2]Staph Membrane Lab, Infectious Diseases, LABiomed, Harbor- UCLA Medical Center, Torrance, USA.
[€]Both authors contributed equally to this work.
*Corresponding author email: prasadtulika@hotmail.com; prasadtulika@mail.jnu.ac.in

The opportunistic dimorphic fungal pathogen, *Candida albicans* is among the top five most common causes of global nosocomial infections leading to almost 40% mortality and morbidity in immunocompromised patients. Increasing incidence of multidrug resistant strains has emerged as a significant threat to the treatment of *Candidiasis*.
The fundamental physical properties of the cell membrane are strongly linked with their biological functions. Membrane provides a permeability barrier to the entry of biomolecules across the cell. It is extremely interesting to investigate the key role played by membrane fluidity in regulating the drug resistance observed in the clinical microbial strains. This study has been carried out to characterize the changes in the membrane fluidity in different phases of growth which includes the lag phase, log phase and stationary phase of cells in an isogenic pair of clinical isolate (GU4-sensitive; GU5-resistant) of *Candida albicans*. Microviscosity of membrane has been measured in this study using fluorescence polarization and the probe, 1, 6-Diphenyl 1, 3, 5-hexatriene (DPH). The reciprocal of fluidity (Micro-viscosity) is the measure of fractional resistance to rotational and translational motion of molecule. With changes in growth phase from lag to log phase, the resistant isolate GU5 was found to attain a more rigid membrane than the sensitive isolate GU4 but after log phase it appeared that the equilibrium is achieved during stationary phase for membrane fluidity for the two isolates and they attain identical physical state of the membrane. Therefore, increased rigid behavior of membrane may be responsible for reduction in passive drug diffusion for the resistant isolate in log phase and at the same time, it also appeared that both the isolates attained a steady state of equilibrium during the stationary phase adapting to the changes in physical state of the membrane. GU5 showed overexpression of the ABC pump membrane protein CDR1p which may be attributed to the observed differences in their membrane fluidity. Membrane fluidity is a parameter of the physical state of the membrane which largely affects the passive diffusion of the drugs and hence the observed resistance towards drugs. The resistance towards drugs also appeared to be influenced by the host environment, prior exposure to drugs and overexpression of different drug efflux pump proteins. The *Candida* cells appeared to adapt to changes in the environment through modulation of their membrane properties and lipid composition leading to altered membrane fluidity.

Keywords: *Candida albicans*; drug resistance; membrane fluidity; isogenic clinical isolates; ABC pump protein

1. Introduction

The burden of antifungal resistance has emerged as a major concern of global public health. There has been a dramatic increase since 1970s in the number of immunocompromised individuals associated with increased usage of immunosuppressive therapies and since 1980s in the prevalence of fungal diseases due to AIDS [1].*Candida* infections have been reported to significantly affect human health and attribute to high mortality rates [2]. *Candida* spp. are polymorphic, opportunistic fungi normally residing in the human microbiome as benign, commensal organisms accounting for 75% of the population. However, under immunocompromised conditions, their pathogenic potential is triggered to cause infections ranging from superficial to life threatening systemic mycoses [3]. *Candida spp.* especially *Candida albicans* ranks as the fourth most common cause of hospital acquired infection among immunocompromised individuals causing bloodstream infections with mortality rate ranging between 38 to 49 % [4, 5]. Antifungal agents, such as azoles and echinocandins have played an important role in the treatment of *Candida* infections [1, 6]. In recent years, efforts have been made to overcome emergence of drug resistant fungi by using drug combinations but high cost and serious side effects have limited such combinatorial therapy [6].

Availability of fewer antifungals puts a constraint to the clinicians' therapeutics which gets further narrowed down by the emergence of drug resistance/ azole resistant strains [7, 8]. Drug tolerance may be an immediate manifestation of response of the organism to drugs leading to subsequent development of drug resistance [8]. Acquisition of multi drug resistance (MDR) may be due to multitude of factors as shown in Fig. 1 which includes i) drug target alteration (by point mutation, overexpression and gene amplification), ii) modification and degradation of drug, iii) increased drug extrusion (by overexpression of drug efflux pumps) and/or iv) decreased drug import or altered intracellular drug accumulation (by altered physical state of the membrane and altered permeability) [7, 8]. Membrane functions may be influenced by factors such as lipid composition, membrane order, lipid asymmetry and glycosylation. Clinical drug resistance is thus multifactorial and may be attributed to a combination of factors related to the host environment, antifungal compounds and pathogen biology. In order to develop an efficient and quick treatment and improved outcome of systemic mycoses, it is essential to have a better understanding of the mechanisms and clinical impact of drug resistance [7, 8].

In addition to the common mechanisms of drug resistance frequently studied, it is important to study changes in structure and function of cell membrane between sensitive and resistant isolates (with prolonged exposure to drugs) for fully elucidating the mechanism of development of drug resistance. This study is an effort to characterize the membrane biophysical state and its possible role in clinical drug resistance for a pair of matched isolates (isogenic) for *Candida albicans*, obtained from the same AIDS patient with recurrent episodes of orophrayngeal *Candidasis* (OPC) and which became fluconazole resistant during therapy [9]. Correlation between resistance and alterations in membrane biophysical properties of the sensitive and resistant *Candida* isolates has been elucidated in this study by using environment-sensitive fluorescent probe, 1, 6-Diphenyl 1, 3, 5-hexatriene (DPH).

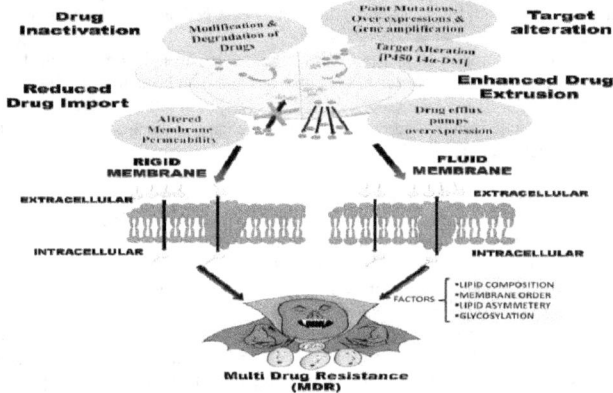

Fig. 1: Molecular mechanisms of fungal multi drug resistance

2. Materials and methods

2.1 Materials, Strains, Media and Growth Conditions

Analytical grade chemicals and HPLC grade solvents used for this study were obtained from HI-Media (Mumbai, India) and Fisher Scientific (Mumbai, India). Lyticase, DPH, tetrahydrofuran (THF) and antifungal drugs viz. fluconazole, itraconazole, ketoconazole, terbinafine were procured from Sigma Chemicals Aldrich (St. Louis, Mo., USA). Table 1 lists the strains used for this study. Cells grown on YEPD (1% Yeast Extract, 2% Peptone, and 2% D-glucose) at 30°C were used for all experiments.

Table 1: Description of strains used for this study

Strain Name	Strain Description
SC5314	Wild type [10]
GU4	Fluconazole sensitive strain with low levels of *CDR1/2* mRNA [9]
GU5	Fluconazole resistant strain with high levels of *CDR1/2* mRNA [9]

2.2 Minimum Inhibitory Concentration (MIC) determination

MIC was determined by broth micro-dilution method as described earlier [11, 12] in accordance with the recommendations of the Clinical and Laboratory (CLSI), formerly National Committee for the Clinical Laboratory Standards (NCCLS).A drug free control was also included. The following concentrations of stock solution of drugs along with their respective solvents (given in parentheses) were used: fluconazole 5mg/ml (DMSO); ketoconazole 1mg/ml (DMSO); itraconazole 1mg/ml (DMSO) and terbinafine 1mg/ml (Ethanol).

2.3 Assessing the physical state of the membrane

Membrane fluidity of cell membrane was monitored using fluorescent probe, DPH as described earlier [11, 13, 14] with slight modifications. Spheroplasts were prepared using lyticase enzyme at 30°C for 3-4 hrs. The spheroplasts were resuspended in labelling buffer pH 7.4 (0.6M Sorbitol, 10mM $MgSO_4$ and 20mM Tris-Cl) and incubated with 2μm DPH for 1hr at 30°C. Fluorescence polarization (p) was calculated as follows on Perkin-Elmer LS55 spectrofluorimeter (excitation 360 nm, emission 426 nm, slit size for both excitation and emission 10 nm) [15]:

$$p = \frac{I_{VV} - (I_{VH} \times G)}{I_{VV} + (I_{VH} \times G)}$$

Where, I_{VV} = Corrected fluorescence intensity obtained with excitation by vertically polarized light and emission detected by analyzer oriented vertically to the direction of polarized excitation light, I_{VH} = Corrected fluorescence intensity obtained with excitation by vertically polarized light and emission detected by analyzer oriented horizontal to the direction of polarized excitation light. *Grating Factor G*, correction for optical components is calculated as I_{HV}/I_{HH} where subscripts HV and HH indicate corrected fluorescence intensity values obtained with horizontal-vertical orientation and horizontal-horizontal orientation for the polarizer and analyzer in that order respectively.

The time resolved fluorescence decay for DPH (excitation 360 nm and emission at 426 nm) was monitored for the above labelled cells in Edinburgh FL920 Fluorescence Life Time Spectrometer as described elsewhere [16].

2.4 Study of ultrastructure of the cells using Transmission Electron Microscopy (TEM)

TEM was used to study the cellular ultrastructure of the clinical isolates. YEPD grown cells were harvested, PBS washed and fixed with 2.5% glutaraldehyde for 2-3 hours at room temperature (RT) and 4-5 hours at 4°C. Cells were then washed 3-4 times with PBS and dehydrated with graded acetone, cleared with toluene and then infiltrated with toluene and araldite mixture at RT. Samples were put overnight in pure araldite at 50°C and then embedded in 1.5 ml eppendorf tube at 60°C. Semi thin and ultrathin sections cut with ultra-microtome (Ultra-microtome Leica EM UC6) were taken on 3.05 mm diameter and 200 mesh copper grid, stained with uranyl acetate and subsequently observed at 120 kV under TEM (Model name JEOL2100F with Accelerating Voltage 80kV-200kV and Magnification 50 to 1,500,000X).

3. Results & Discussion

Isogenic/ matched pair of clinical isolates (GU4-sensitive and GU5-resistant) collected from same AIDS patient, which became fluconazole resistant during therapy were used for this study along with wild type SC5314. MIC_{90} (μg/ml) (Table 2) confirmed the MDR phenotype for GU5. Highest MIC was observed for GU5 for fluconazole, ketoconazole, itraconazole and terbinafine whereas SC5314 and GU4 showed sensitivity towards these drugs. GU4 showed a relatively higher MIC for all drugs than SC5314 which may be largely due to their prior exposure to drug and complex host factors in the human patient.

Table 2: MIC_{90} for different drugs for Wild type (SC5314) and clinical isolates (GU4 and GU5)

Strain Name	MIC $_{90}$ in μg/ml			
	Fluconzaole	Itraconazole	Ketoconazole	Terbinafine
SC5314	3.125	0.015	0.062	3.06
GU4	6.25	0.031	0.125	6.125
GU5	>200	1	>2	12.25

Cell envelope serves as a barrier for entry or exit of biomolecules across the cells. Many cellular changes with different structure, function and mechanism may accompany the drug resistance phenotype which may be intrinsic (inherent) or acquired. As an effort to study the multitude of factors involved in acquiring clinical resistance, membrane biophysical properties for the sensitive and resistant isolates were analyzed in this study at different phases of growth using fluorescence polarization. *Candida* cells showed a typical sigmoid growth pattern (Fig. 2A) with early lag phase till 6 hours, lag phase till 10 hours, early log phase at 12 hours, log/ exponential phase between 14-18 hours and approaching stationary phase 20 hours onwards.

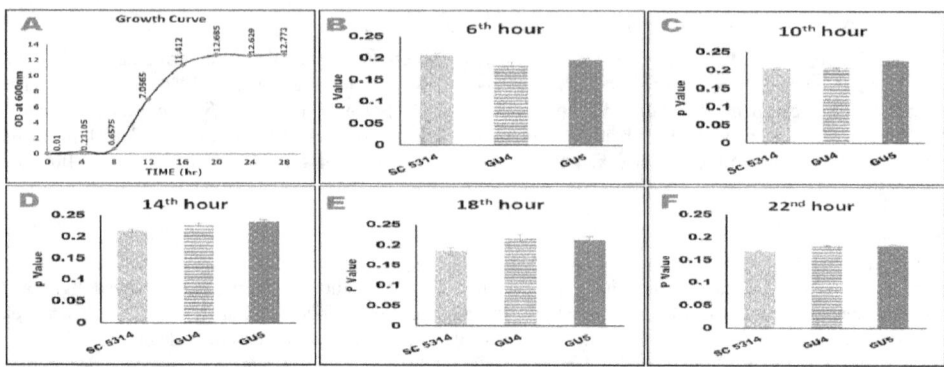

Fig. 2: (A) Growth curve of *Candida* cells. **(B-F)** Mean fluorescence polarization "p" values (inversely proportional to membrane fluidity) ± the standard deviation of the mean of the three sets of experiments, of wild type (SC5314) and clinical isolates (GU4 and GU5) at different hours of growth.

Sterol rich membrane microdomains have been reported to exist in eukaryotes and play a crucial role in membrane structure, organization and function. Sterols and specific lipids have been found to be non-randomly distributed in such domains of biological membranes [17]. Fluorescence being sensitive to the cellular physicochemical environment, DPH was used to study organizational dynamics of the hydrophobic interior of the membrane in the sensitive and resistant isolates. Statistically significant differences were observed between them at different phases of growth by steady state fluorescence polarization and time resolved fluorescence measurements (Fig. 2B-F and Table 3). Packing of sterol and fatty acyl chain is likely to affect the diffusion rate of membrane embedded probes, thus affecting fluorescence polarization values. Fluorescence polarization is inversely proportional to membrane fluidity and increase in fluorescence polarization values are typically due to a reduction in rotational mobility of the fluorophore, which is influenced by dense packing of membrane. GU5 showed increase in membrane rigidity than its sensitive pair GU4 with increase in hours of growth and fluorescence polarization values increased in GU5 by 4.43% (6 hours), 8.44% (10 hours) and 3.77% (14 hours). No significant differences were observed between the two isolates at 18 hours and 22 hours of growth and fluorescence polarization values in GU5 decreased by only 2.06% (18 hours) and increased by 0.5 % (22 hours). Wild type SC5315 in absence of prior exposure to host factors showed different behaviour than the clinical pair i.e. higher rigidity at 6 hour of growth and from 10 hour onwards revealed higher fluidity than the other two clinical isolates. The observed results may be due to differences in the levels of compaction of cell membrane between sensitive and resistant isolates and differences in lateral organization on the plane of membrane contributed by ergosterol, sphingolipids and overexpressed drug efflux pump proteins.

Fluorescence lifetime is a reliable indicator of polarity changes in local environment of the DPH. Life time of DPH has been found to be reduced in presence of water in its immediate environment [17]. The mean fluorescent lifetimes (Table 3) were calculated from the fitting of tri-exponential fluorescent decay curves [16]. The data revealed that DPH lifetime for the resistant isolate is always lower than the sensitive isolate and the decay for DPH becomes faster with increase in time of growth from 6 to 22 hours. The sensitive isolate shows gradual lowering of decay time of DPH from 6 hours to 18 hour of growth and increases again at 22 hours of growth in sensitive isolate. Increase in water penetration might have increased the polarity of the environment resulting in shortening of DPH lifetime. Previous reports mention that ergosterol beyond a certain concentration do not influence dynamics and membrane order [17]. Water penetration may be more due to lesser rigidifying effect of ergosterol as revealed by fluorescence polarization. These decay patterns provided the observed changes in the cellular microenvironment of the respective membrane during different phases of growth.

Table 3: Lifetime decay for DPH in nanoseconds (ns) at different hours of growth for the clinical isolates (GU4 and GU5)

Strains Name	6th Hour (ns)	10th Hour (ns)	14th hour (ns)	18th Hour (ns)	22nd Hour (ns)
GU4	2.8	1.9	0.94	0.75	1.12
GU5	2.24	1	0.68	0.57	0.73

It is evident that with changes in growth phase from lag to log phase, the resistant isolate GU5 attained a more rigid membrane than the sensitive isolate GU4 but after log phase it appeared that equilibrium was achieved for membrane fluidity. This data indicated that membranes of the clinically resistant isolate (GU5) exhibited rigidity favouring the development of drug resistance. Rigid behaviour of membrane is likely to show reduction in passive drug diffusion. At the same time, all isolates appeared to attain a steady state of equilibrium during the stationary phase adapting to the changes in physical state of the membrane. It may be noted that GU5 showed overexpression of the ABC pump membrane proteins CDR1/2p [9].

There are reports suggesting involvement of ultrastructural changes in the cell envelope in development of drug resistance which includes increase in cell wall thickness [18]. Although the changes observed in membrane order were dependent on the cellular microenvironment, thickening of cell wall was consistently observed with increase in clinical resistance. The cell wall thickness was in the order: wild type SC5314 (74.6 nm) < sensitive GU4 (90.3 nm) < resistant GU5 (117nm). Thickness of cell wall of each isolate was measured from outer cell membrane border to the outer cell wall border.

There are various factors determining the membrane structure and influencing binding, transport of different ionic/ molecular species which include membrane phase state, hydration, electrostatics, dynamics of the constituent molecules etc [19]. Order of the lipids is a parameter that defines different phases of membrane state: i.e. gel (Lb), Liquid order (Lo) and fluidity (La) [19]. The liquid order (Lo) which is responsible for the membrane microdomains presents a higher level of lipid order and microviscosity compared to fluid phase [19]. The resistance towards drugs is also influenced by the host environment, prior exposure to drugs and over expression of different drug efflux pump proteins.

Membrane properties in this study have been defined at different phases of growth and the physical state of the membrane are reflected in the spectroscopic response of the fluorescence probes used. The clinical sensitive isolate showed difference in membrane order as compared to the laboratory isolate due to their prior exposure to

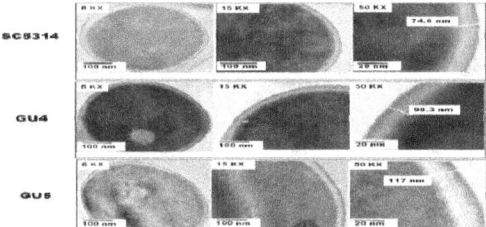

Fig. 3: TEM images of cell envelope of isolates.

variety of host factors and to drugs. The resistant isolate due to over expression of different membrane efflux pump proteins also exhibited different physical state of the membrane. Hence it appears that there are sweet spots in membrane which gradually helps the membrane to acquire a steady state of equilibrium towards the stationary phase of growth. Cell wall thickness observed may be associated to altered cell wall synthesis or may be a non-specific secondary adaptation response linked to change in cell membrane fluidity and/ or function.

This study revealed that membrane fluidity is a parameter of the physical state of the membrane which largely contributes to the passive drug diffusion and observed resistance towards drugs. It of course, merits further investigation to determine if observed differences in cell wall and cell membrane fluidity are strain specific or generalized observations.

Acknowledgements: This work has been supported by grants to TP from Department of Biotechnology (BT/PR5110/MED/29/497/2012 and BT/BI/12/045/2008), University Grants Commission UPE II scheme and Department of Science and Technology (DST-PURSE), India. We acknowledge Joachim Morschhauser, University of Wurzburg for generously providing clinical isolates and expert help of Sobhan Sen, Jawaharlal Nehru University (JNU) for time resolved data analysis. JNU and AIRF for providing infrastructural support and Department of Biotechnology (DBT), India for the award of Senior Research Fellowship to VSR and Junior Research Fellowship to RS are gratefully acknowledged.

References

[1] Odds FC, Brown AJP and Gow NAR. Antifungal agents: mechanisms of action. Trends Microbiol. 2003; 11(6):272-279.
[2] Weig M and Brown AJP. Genomics and the development of new diagnostics and anti *Candida* drugs. Trends Microbiol. 2007; 15(7):310-317.
[3] Mayer FL, Wilson D and Hube B. *Candida albicans* pathogenicity mechanisms. Virulence. 2013; 4(2):119–128.
[4] Liu S, Hou Y, Chen X, Gao Y, Li H, Sun S. Combination of fluconazole with non-antifungal agents: A promising approach to cope with resistant *Candida albicans* infections and insight into new antifungal agent discovery. Int J Antimicrob Agents.2014; 43(5):395–402.
[5] Arendrup MC, Dzajic E, Jensen RH, Johansen HK, Kjaeldgaard P, Knudsen JD, Kristensen L, Leitz C, Lemming LE, Nielsen L, Olesen B, Rosenvinge FS, Røder BL and Schønheyder HC. Epidemiological changes with potential implication for antifungal prescription recommendations for fungaemia: data from a nationwide fungaemia surveillance programme. Clin. Microbiol Infect. 2013; 19(8):343–353.

[6] Cannon RD, Lamping E, Holmes A. R, Niimi K, Tanabe K, Niimi M and Monk BC. *Candida albicans* drug resistance another way to cope with stress. Microbiology. 2007; 153(10):3211–3217.

[7] Kanafani ZA and Perfect JR. Resistance to Antifungal Agents: Mechanisms and Clinical Impact. Clin. Infect Dis. 2008; 46(1):120-128.

[8] Prasad T, Sethumadhavan S and Fatima Z. Altered Ergosterol biosynthetic pathway - an alternate multidrug resistance mechanism independent of drug efflux pump in human pathogenic fungi *C. albicans*. Science against microbial pathogens: communicating current research and technological advances. Formatex Microbiology series. 2011; 3:757-768.

[9] Franz R, Ruhnke M and Morschhauser J. Molecular aspects of fluconazole resistance development in *Candida albicans*. Mycoses. 1999; 42(7-8):453–458.

[10] Fonzi WA and Irwin MY. Isogenic Strain Construction and Gene Mapping in *Candida albicans*. Genetics. 1993; 134(3):717-728.

[11] Prasad T, Chandra A, Mukhopadhyay CK and Prasad R.Unexpected Link between Iron and Drug Resistance of *Candida spp.*: Iron Depletion Enhances Membrane Fluidity and Drug Diffusion, Leading to Drug-Susceptible cells. Antimicrob. Agents Chemother. 2006; 50(11):3597–3606.

[12] Wayne, PA. Reference Method for Broth Dilution Antifungal Susceptibility Testing of Yeasts. Approved Standard M27-A3, third ed. Clinical and Laboratory Standards Institute. CLSI. 2008.

[13] Prasad T, Hameed S, Manoharlal R, Biswas S, Mukhopadhyay CK, Goswami SK and Prasad R. Morphogenic regulator *EFG1*affects the drug susceptibilities of pathogenic *Candida albicans*. FEMS Yeast Res.2010; 10(5):587–596.

[14] Mukhopadhyay K, Prasad T, Saini P, Pucadyil TJ, Chattopadhyay A, and Prasad R. Membrane Sphingolipid-Ergosterol Interactions Are Important Determinants of Multidrug Resistance in *Candida albicans*.Antimicrob. Agents Chemother.2004; 48(5):1778–1787.

[15] Shinitzky M and Barenholz Y. Fluidity parameters determined by fluorescence polarization. Biochim. Biophys. Acta. 1978; 515:367–394.

[16] Verma SD, Pal N, Singh MK and Sen S. Probe position-dependent counterion dynamics in DNA: Comparison of time-resolved Stokes shift of groove-bound to bsae stacked probes in the presence of different monovalent counterions. J PhysChemLett. 2012; 3:2621-2626.

[17] Arora A, Raghuraman H and Chattopadhyay A. Influence of cholesterol and ergosterol on membrane dynamics: a fluorescence approach. BiochemBiophys Res Commun. 2004; 318:920-926.

[18] Mishra NN, Bayer SA, Tran TT, Shamoo Y, Mileykovskaya E, Dowhan W, Guan Z and Arias CA. Daptomycin resistance in Enterococci is associated with distint alteration of cell membrane phospholipids content. PLoS One. 2012; 7(8):e43958:1-10.

[19] Klymchenko AS and Kreder R. Fluorescent probes for lipid rafts: from model membranes to living cells. Chem. Biol. 2014; 21:97-113.

A protocol for screening protein-protein interaction inhibitors with the "Two phages" Two Hybrid assay

L. Grenga[*,1] and **P. Ghelardini**[2]

[1] Laboratory of General Microbiology, Biology Department, University of Rome Tor Vergata, via della Ricerca Scientifica 1 00133 Rome, Italy

[2] Institute of Biology, Molecular Medicine and Nanotechnology of CNR c/o Biology Department, University of Rome Tor Vergata, Rome, Italy

*Corresponding author: e-mail: lucia.grenga@uniroma2.it , Phone: +39 06 72594217

Bacterial cell division is an essential process, interesting for two main reasons: a basic biological knowledge and the possibility to exploit the cell division proteins as primary targets for novel broad-spectrum antibacterial drugs. The "Two phages" two hybrid assay (THA) already used to depict the *E. coli* and *S. pneumoniae* cell division interactome, could constitute a useful tool to select small molecules interfering with the interactions among the division proteins. These molecules, binding to residues involved in protein interactions, on the proteins contact surfaces, could provide both a complementary and more flexible approach for studies on molecular mechanism of cell division and, as important applicative consequence, they potentially constitute a novel class of antimicrobial agents with a very low risk of resistance. At this regard, we set the assay, described in this paper, for the screening of small molecules (or peptides/peptidomimetics) to identify protein-protein interaction inhibitors. This assay, based on the two phages THA, was validated using 3'-(2-phenyl-1H-indol-3-yl)-[1,1'-biphenyl]-3-carboxylic acid, that is known for interfering with the interaction between the *E. coli* division proteins FtsZ and ZipA.

Keywords β-galactosidase activity; Protein-protein interaction; Inhibitors; Lambdoid phage repressors; Small molecules; Two hybrid assay

1. Introduction

Proteins are the main effectors of cellular processes. In fact, most of the cell functions are the result of complex machineries where proteins are organized in dense interaction networks, either stably assembled in multi-protein cellular 'machines' or transiently interacting with one another in a signal transduction cascade.

Among the numerous *in vivo* and *in vitro* approaches described to study the protein interactions, the bacterial "Two phages" two hybrid assay (THA) [1] was successfully used to study the *E. coli* and *S. pneumoniae* division interactome [2], clarifying some of the molecular aspects of prokaryotic divisome formation. The study of bacterial division interaction network shows an important applicative fallout since protein–protein interactions (PPIs) have emerged as promising drug targets. Inhibitors of these interactions could account for the purpose of the modern pharmaceutical research focused on the identification of novel antibacterial agents able to play down the rising of bacterial resistance that, still represents the main problem of antibiotic therapy.

Small molecules that bind to specific residues on the protein contact surfaces [3] could constitute a novel class of antimicrobial agents with a very low risk of resistance. As a matter of fact, impairing the interaction between two proteins A and B forming a heterodimer A-B, essential for the bacterial survival, will be lethal. Bacterial mutants, antiA-B resistant, need a mutation at the interface between antiA-B and A or B protein. These mutants will be also lethal, since the protein, mutated in the interaction site, will be not able to interact with its wild type partner. Only the double mutants, simultaneously mutated in both the two partner proteins, will be selected. This kind of mutants is much less frequent compared to the mutants in the A or B domains, recognized by the old style antibiotics, resulting in a very low rate of bacterial resistance.

At this regard, a screening assay based on the "Two phages" THA, characterized by feasibility, reproducibility and low cost, could constitute a useful tool to identify small molecules that, interfering with the protein interactions, could therapeutically modulate the bacterial cell division.

2. Materials and Methods

2.1 Media and chemicals

LB broth for bacterial culture and plating and SM (salt solution) for bacteria dilutions were as described by Miller [8]. The antibiotics, ampicillin (50 µg ml⁻¹), tetracycline (40 µg ml⁻¹) and kanamycin (30 µg ml⁻¹), DMSO and Polymyxin B nonapeptide (PMBN) were purchased by Sigma.

2.2 Bacterial strains and plasmids

Bacterial strains, *E. coli* K-12 derivatives, and plasmids used in this work are listed in Table 1.

2.3 General microbiological and recombinant DNA techniques

Standard microbiological techniques were as described by Miller [4]. Standard procedures were used for small-scale plasmid preparations, agarose gel electrophoresis and bacterial transformation [5]. PCR was carried out using the *Taq* DNA polymerase kit (Promega), according to the recommendations of the manufacturer.

2.4 Assay conditions

The assay derives from the "Two phages" THA, already described [2]. The *E. coli* 7118SB bacterial strain was transformed with the two plasmids pcI$_{P22}$ and pcI$_{434}$ containing the genes coding for the two interacting proteins, against which small molecules as inhibitors were looking for, were under the control of p$_{LAC}$ and p$_{ARA}$ respectively (Figure 1). Clones were selected on LB plates supplemented with the opportune antibiotics and 1% glucose to inhibit the expression from p$_{LAC}$. Clones were grown in LB in 2 ml wells plate at 37°C to OD$_{600}$ 0.3 (usually it takes about 90 minutes). The culture was then diluted to an OD$_{600}$ of about 0.1 and distributed in aliquots of 200 μl of LB supplemented with 2.5 μg/ml PMBN and IPTG 1×10^{-4} M, to induce the expression of one of the two proteins under investigation, in 96 wells plate. At this time (T$_0$ of the experiment) the inhibitor in DMSO was added to each aliquot. After incubation at 37°C for 30 minutes, L-arabinose 0.2%, to induce the expression of the partner protein, was added and the plate was incubated at 37°C to T=180 minutes. The OD$_{600}$ was determined on aliquots of 100 μl, withdrawn from each well, and the β-galactosidase activity was tested on 50 μl of the remaining culture, as described by Miller [3], lysing the cells with SDS and chloroform. The Miller units of β-galactosidase produced from each culture were normalized with that produced by the 7118SB strain without or harbouring only one plasmid. In the "Two phages" THA, the interaction, and its strength, was calculated from the amount of β-galactosidase produced by the 7118SB strain, harbouring the two plasmids, normalized on that produced by the same strain harbouring only one of the two plasmids. A residual activity less than 50% is indicative of protein interactions [6].

If the interaction between the two proteins, under investigation, was inhibited by an interfering small molecule, the residual β-galactosidase activity will increase from less than 50% up to the 100% value of the reporter strain.

In all the experiments described in this paper, the reported β–galactosidase values were the means of at least four independent determinations where the standard deviation did not exceed ± 4%.

3. Results and discussions

3.1 Setting the assay

Various parameters have been taken into account in order to set a protocol for screening small molecules impairing PPIs with the "Two phages" THA.

a) Bacterial growth in the assay conditions

The assay was set to be performed in 96 well plates. The bacterial growth of strain 7118SB and of its derivatives harbouring the two plasmids pcI$_{434}$ and pcI$_{P22}$, was examined in a final volume of 200 μl of LB medium, without shaking. In the assay condition, 7118SB strain grew with a generation time of 45 minutes that became 50 and 60 minutes for its derivatives, depending on the proteins cloned in the two plasmids, when their expression was induced with 1×10^{-4} M IPTG and 0.2% L-arabinose. Growth saturation was reached early, at OD$_{600} \sim 0.4$ for strain without or harbouring only one plasmid. When both the two proteins forming the chimeric repressor were expressed in the bacterial strain, the growth was blocked at OD$_{600}$ 0.15-0.3, again depending from the expressed proteins. To perform the assay in exponential growth conditions, a pre-culture of the 7118SB strain containing either only one or both plasmids was grown at 37°C in LB to OD$_{600}$ 0.3, then diluted 3 times in LB supplemented with the inducers and let grow for 3 hours. During this period the growth as well as the β-galactosidase synthesis remained exponential (data not shown).

b) Evaluation of both cell wall permeation and concentration of the DMSO on the bacterial growth.

The ability to permeate the *E. coli* bacterial cell, to allow the entry of the small molecules, was tested by evaluating the minimal inhibitory concentration (MIC) of novobiocin in the presence or absence of various amount of Polymyxin B nonapeptide (PMBN), a cationic cyclic peptide derived from the antibacterial peptide Polymyxin B, that specifically increases the permeability of the outer membrane of Gram-negative bacteria toward hydrophobic antibiotics, testing at the same time its effect on cell viability. As expected [7], the MIC of novobiocin, which was 50 µg/ml in the absence of PMBN, became 6.2 µg/ml in the presence of 2.5 µg/ml of PMBN and this value remained constant increasing the PMBN concentration up to 5 µg/ml and 10 µg/ml. The bacteria growth rate was only slightly reduced with a PMBN concentration of 2.5 µg/ml (<10% increase of the generation time after 3 hours of incubation) whereas with 5 µg/ml the increase of generation time was more than 30%.

Since compounds to be tested should be re-suspended in DMSO, its effects on bacterial growth were also evaluated adding various concentrations to an exponentially growing culture, in the assay conditions. After 3 hours of incubation, the bacterial growth, unaffected in the presence of DMSO ranging from 0 to 2%, decreases of 60% with 5% and of 90% with 10% DMSO.

In conclusion, the assay conditions foresaw the presence of at least 2.5 µg/ml of PMBN to allow the entry of the compounds that should be re-suspended in DMSO in such a way that its final concentration in the bacterial culture should not exceed 2%.

3.2 Effect of 3'-(2-phenyl-1H-indol-3-yl)-[1,1'-biphenyl]-3-carboxylic acid on FtsZ-ZipA interaction

Sutherland et al., [8] described some compounds that can interfere with the interaction between the *E. coli* FtsZ and ZipA proteins, identifying the domain constituted by the last 15 residues of FtsZ as the target of their action. In order to validate our assay for selecting PPIs inhibitors, we studied, in our conditions, the inhibitory action on FtsZ-ZipA interaction of one of these compounds, the 3'-(2-phenyl-1H-indol-3-yl)-[1,1'-biphenyl]-3-carboxylic acid (C27H19NO2), purposely synthetized by "Coliseum Combinatorial Chemistry Centre for Technology (C4T)" (Rome, Italy).

To study the effects of C27H19NO2 on the bacterial growth, we measured its MIC in strain 7118SB. According with Sutherland et al., [8], it was about 300 µM, whereas in the assay conditions, i.e. in the presence of of 2.5 µg/ml PMBN, the MIC was reduced to 20 µM.

a) Study of FtsZ-ZipA interaction in the assay condition

As a first step, we checked whether the assay conditions were suitable to reveal the FtsZ-ZipA interaction. Two sets of plasmid pairs, formed by $pcI_{434}ftsZ$ and $pcI_{P22}zipA$ and $pcI_{434}ftsZ_{15Cter}$ (coding for the FtsZ derivative domain of 15 residues) and $pcI_{P22}zipA$, respectively, were used. In the control strains 7118SB without plasmids, the β-galactosidase produced was on the order of 2000 Miller Units and, as expected, this value did not change in its derivative strain harbouring only the $pcI_{434}ftsZ$ plasmid. The β-galactosidase values obtained in the assay, using the strain harbouring both plasmids, were normalized on this value, which constituted the 100% of residual activity (Figure 2). Results indicated that the interaction ability increases as a function of time from 41% of residual activity after 3 hours of induction to 27% after 5 hours, depending on the amount of proteins expressed in the cell upon induction. As shown in Figure 2, both the whole FtsZ protein and its domain $FtsZ15_{Cter}$ behave in the same way as far as the interaction with ZipA is concerned.

b) Inhibition of FtsZ-ZipA interaction with C27H19NO2

Various parameters were taken into account to study the FtsZ-ZipA interaction inhibition with C27H19NO2 by means the "Two phages" THA: (i) the interaction target analysing the inhibitory with both the whole FtsZ protein and its derivative of 15 residues, (ii) kinetics of inhibition, studied at 0, 180 and 300 minutes after the C27H19NO2 addition, with two concentration of the compound, 100 and 300 µM (iii) effect of inhibitor concentration. The kinetics of inhibition was and assays were performed with 0, 10, 25, 50, 75, 100 and 300 µM of C27H19NO2, to determine its minimal and the optimal concentrations to inhibit the interaction. Lastly, (iv) specificity of the inhibitory effect was examined.

(i) As expected, from the data of Sutherland et al., [8], no differences in interaction inhibition were observed performing the assay with both FtsZ and $FtsZ_{15Cter}$ (Figure 3b). (ii) Interaction inhibition due to C27H19NO2 is maximal at T=180 minutes with 100 µM of the compound (Figure 3a). The reduction of the inhibitory effect over this concentration could be due to a precipitation of the compound that is poorly soluble at 300 µM. In every case, it should be taken into account that the inhibitory effect was the result of counterbalance between the C27H19NO2 action and the amount of its target protein interacting with its partner in the cell.

(iii) In Figure 3c, the results of the inhibitory effect as a function of the compound concentration were reported. The maximal activity of C27H19NO2 was observed between 25 and 50 µM where the residual activity of β-galactosidase was of the same order of the control. These concentrations are slightly superior to the minimal inhibitory concentration (MIC), which was 20 µM, in the same experimental conditions.

(iv) Lastly, we analyzed the specificity of C27H19NO2 action studying its effects on three pairs of couples of interacting proteins. Two of them belonged to the *E. coli* division interactome and the third one was formed by the C-terminal part of the 434 phage repressor fused in frame with the N-terminal portions of phages 434 and P22 repressors, respectively, to originate a functional chimeric 434 repressor. The two divisome couples were FtsZ-FtsA, whose interaction is localized at C-terminal of FtsZ protein, involving the same site of FtsZ-ZipA interaction. Indeed FtsA and ZipA compete for the same site on FtsZ [9]. The second couple was formed by FtsQ and FtsI. This couple of proteins was chosen since none of them interacts with FtsZ [1]. The results of the experiments, reported in Figure 4, showed that the interaction between FtsA and FtsZ was inhibited at the same extent as FtsZ-ZipA. This result could somehow be expected since the same domain of FtsZ is involved in both FtsA and ZipA interactions.

On the other hand, when the 7118SB strain harboured the plasmid pairs coding for either FtsI and FtsQ or for the two subunits of 434 phage repressor, the residual β-galactosidase activity remained almost constant and comparable to that obtained in the absence of inhibitor, despite the presence of increasing amounts of C27H19NO2, highlighting its specific action only on the FtsZ C-terminal domain.

In conclusion, this assay can be used to test ligands with a wide range of size (from 15 to about 400 residues, in the reported case) and proved to be selective and reproducible. Due to its feasibility and operating speed it can be easily adapted for high-throughput drug discovery efforts in the screening of small molecules inhibitors of protein-protein interactions.

Table 1 Bacterial strains and plasmids used in this work

Bacterial strain	Relevant genotype	Source
71/18	SupE thy Δ(lac-proAB9 F' [proAB$^+$ lacI^q lacZDM15]	[10]
7118SB		[11]
	71/18 *glpT*::O-P434/P22 lacZ	

Plasmids		
$p_{ARA}cI_{434}$	$P_{LAC}cI_{434}$ derivative harbouring the *araC* p_{ARA} region	This work
$p_{ARA}cI_{434}ftsZ$	$P_{LAC}cI_{434}$ derivative harbouring the *E. coli ftsZ* gene	This work
$p_{ARA}cI_{434}ftsZ_{15Cter}$	$P_{LAC}cI_{434}$ derivative harbouring the 15 residues at C-ter of *E. coli ftsZ* gene	This work
$p_{ARA}cI_{434}ftsI$	$P_{LAC}cI_{434}$ derivative harbouring the *E. coli ftsI* gene	This work
$p_{ARA}cI_{434}cI_{434Cter}$	$P_{LAC}cI_{434}$ derivative harbouring the C-ter of 434 repressor	This work
$p_{LAC}cI_{P22}zipA$	pcI_{P22} derivative harbouring the *E. coli zipA* gene	Our Lab
$p_{LAC}cI_{P22}ftsA$	pcI_{P22} derivative harbouring the *E. coli ftsA* gene	Our Lab
$p_{LAC}cI_{P22}ftsQ$	pcI_{P22} derivative harbouring the *E. coli ftsQ* gene	Our Lab
$p_{LAC}cI_{P22}cI_{434Cter}$	pcI_{P22} derivative harbouring the C-ter of 434 repressor	Our Lab

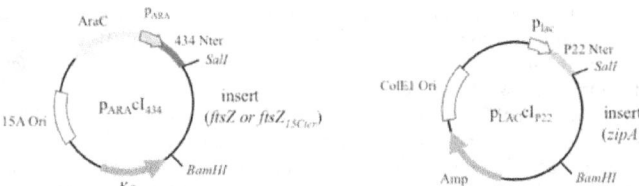

Fig. 1 *Maps of plasmids $p_{ARA}cI_{434}ftsZ$ and $p_{LAC}cI_{P22}zipA$.* Schematic representation of plasmids $p_{ARA}cI_{434}ftsZ$ and $p_{LAC}cI_{P22}zipA$.

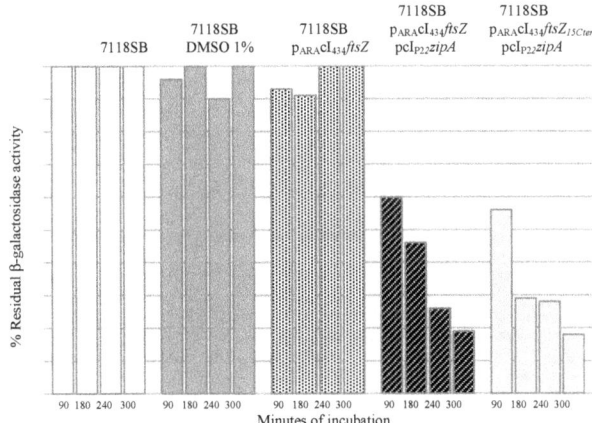

Fig. 2 *Protein interaction in the assay conditions.* Residual β-galactosidase activity measured as function of time in *E. coli* 7118SB strain, supplemented or not with 1% DMSO, and its derivatives containing either the $p_{ARA}cI_{434}ftsZ$ and $p_{LAC}cI_{P22}zipA$ plasmids or $p_{ARA}cI_{434}ftsZ_{15Cter}$ and $p_{LAC}cI_{P22}zipA$.

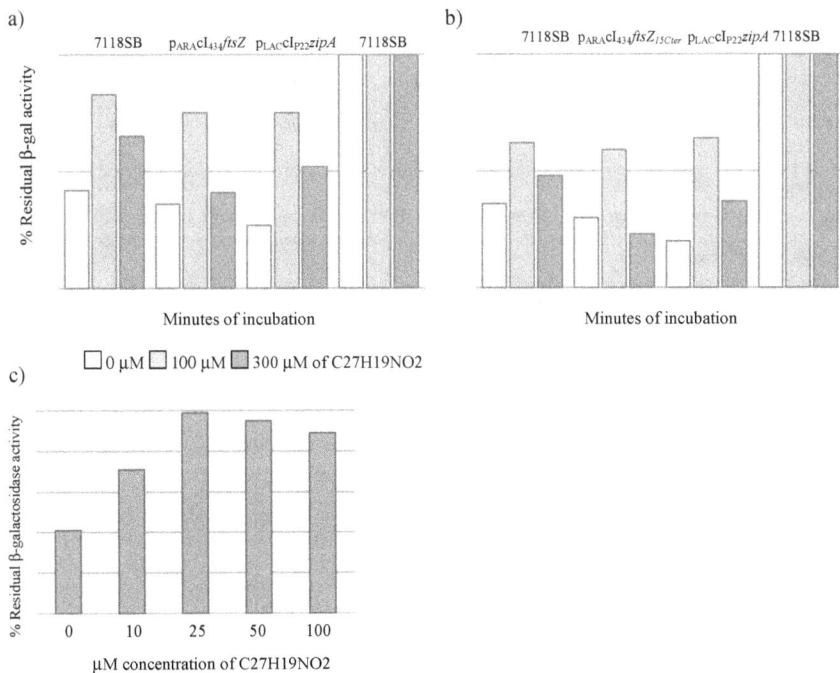

Fig. 3 *Effect of C27H19NO2 on FtsZ (or FtsZ15$_{Cter}$) ZipA interaction.* a) Residual β-galactosidase activity produced by 7118SB strain harbouring $p_{ARA}cI_{434}ftsZ$ and $p_{LAC}cI_{P22}zipA$ plasmids or $p_{ARA}cI_{434}ftsZ_{15Cter}$ and b) $p_{LAC}cI_{P22}zipA$ as a function of time of incubation with various concentrations of C27H19NO2. c) Residual β-galactosidase activity produced by 7118SB strain harbouring $p_{ARA}cI_{434}ftsZ$ and $p_{LAC}cI_{P22}zipA$ plasmids after 3 hours of incubation with various concentrations of C27H19NO2. This figure is also representative of the behaviour of 7118SB strain harbouring $p_{ARA}cI_{434}ftsZ_{15Cter}$ and $p_{LAC}cI_{P22}zipA$.

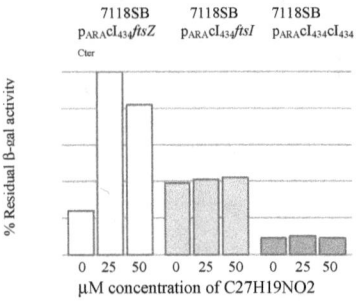

Fig. 4 *Specificity of interaction inhibition with C27H19NO2.* Residual β-galactosidase activity produced by 7118SB strain harbouring various pairs of plasmids coding for interacting proteins (namely: $p_{ARA}cI_{434}ftsZ$ $p_{LAC}cI_{P22}ftsA$, $p_{ARA}cI_{434}ftsI$ $p_{LAC}cI_{P22}ftsQ$ and $p_{ARA}cI_{434}cI_{434Cter}$ $p_{LAC}cI_{P22}cI_{434Cter}$) after 3 hours of incubation with various concentrations of C27H19NO2.

Acknowledgements   We are most grateful to C4T for the synthesis of C27H19NO2 and to the Microbiology Laboratory of Angelini ACRAF Pharmaceutical for their support in some experiments. This work was financed by a FILAS biotechnology project and a contract with Angelini ACRAF spa.

References

[1]   Di Lallo G, Fagioli M, Barionovi D, Ghelardini P & Paolozzi, L. Use of a two-hybrid assay to study the assembly of a complex multicomponent protein machinery: bacterial septosome differentiation. Microbiology. 2003; 149:3353–3359.

[2]   Maggi S, Massidda O, Luzi G, Fadda D, Paolozzi L, & Ghelardini P. Division protein interaction web: identification of a phylogenetically conserved common interactome between *Streptococcus pneumoniae* and *Escherichia coli.* Microbiology. 2008; 154:3042-3052.

[3]   Wells JA & McClendon CL. Reaching for high-hanging fruit in drug discovery at protein–protein interfaces. Nature. 2007; 450:1001-1009.

[4]   Miller JH. *Experiments in Molecular Genetics.* Cold Spring Harbor, NY: Cold Spring Harbor Laboratory. 1972.

[5]   Sambrook, J., Fritsch, E. F. & Maniatis, T. *Molecular Cloning: a Laboratory Manual.* Cold Spring Harbor, NY: Cold Spring Harbor Laboratory. 1989.

[6]   Fadda D, Santona A, D'Ulisse V, Ghelardini P, Ennas MG, Whalen MB, & Massidda O. Streptococcus pneumoniae DivIVA: localization and interactions in a MinCD-free context. J Bacteriol. 2007; 189:1288–1298.

[7]   Ofek I, Cohen S, Rahmani R, Kabha K, Tamarkin D, Herzig Y, & Rubinstein E. Antibacterial synergism of polymyxin B nonapeptide and hydrophobic antibiotics in experimental gram-negative infections in mice. Antimicrob Agents Chemother. 1994; 38:374-377.

[8]   Sutherland AG, Alvarez J, Ding W, *et al.* Structure-based design of carboxybiphenylindole inhibitors of the ZipA-FtsZ interaction. Org Biomol Chem. 2003; 1:4138-4140.

[9]   Margolin W. The assembly of proteins at the cell division site. In *Molecules in Time and Space: Bacterial Shape, Division and Phylogeny,* Vicente, Tamames and Mingorance Eds. Kluwer Academic/Plenum Publishers. New York. 2004. p. 79-102.

[10] Messing J, Gronenborn B, Müller-Hill B, & Hans Hopschneider P. Filamentous coliphage M13 as a cloning vehicle: insertion of a HindII fragment of the lac regulatory region in M13 replicative form in vitro. Proc. Natl. Acad. Sci. USA. 1977; 74: 3642-3646.

[11] Barbati S, Grenga L, Luzi G, Paolozzi L & Ghelardini P. Prokaryotic division interactome: setup of an assay for protein-protein interaction mutant selection. Research in Microbiology. 2010; 161:118-126.

Adsorption and biodegradation of reactive orange 16 by *Funalia trogii* 200800 in a biofilm reactor using activated carbon as a supporting medium

Yen-Hui Lin[*]

Department of safety, Health and Environmental Engineering, Central Taiwan University of Science and
   Technology, 666 Bu-zih Road, Bei-tun District, Taichung 40601, Taiwan
*Corresponding author: e-mail: yhlin1@ctust.edu.tw, Phone: +886 422391647x6861

A non-steady-state mathematical model system for the kinetics of adsorption and biodegradation by *Funalia trogii* (*F. trogii*) cells on activated carbon was derived. The batch kinetic tests were conducted to determine biokinetic and adsorption parameters. The yield coefficient of *F. trogii* cells obtained from a batch kinetic test was equal to 0.204 mg cell/mg RO16. The maximum specific growth rate (μ_m) of *F. trogii* cells was 0.63 day^{-1}. The maximum specific utilization rate (k) of RO16 was 3.1 mg RO16/mg cell-day. The half-saturation constant of RO16 was 107.3 mg RO16/L. The decay coefficient of *F. trogii* cells was 0.024 day^{-1}. Frundlich isotherm tests were conducted to evaluate the adsorption capacity of activated carbon for RO16. The values for Freundlich isotherm coefficients K_q and n were 0.271 (g/g)(L/mg)$^{1/n}$ and 1.755, respectively. A continuous-flow biofilm reactor using activated carbon as a supporting medium was conducted to evaluate the removal efficiency of RO16. The effluent concentration of RO16 was 0.1-0.2 mg RO16/L. The removal efficiency for RO 16 was approximately 98-99%.

Keywords Adsorption; Biodegradation; Biofilm

1. Introduction

The textile wastewater has a specific characteristic of high temperature, high-alkaline and strong color organic compounds. The color of textile wastewater is various and changeable according to the manufacturing process [1,2]. The treatment of dye wastewater involves chemical and physical methods such as adsorption, coagulation, oxidation, filtration, and ionization radiation. All these methods have different decolorization capabilities, capital cost and operating speed. Among these methods, coagulation and adsorption are the commonly used; however, these created huge amounts of sludge which become a pollutant creating its own disposal problems. Hence, biological processes have received increasing interest as a viable alternative owing to their cost effectiveness, ability to produce less sludge and environmental friendliness [3]. Attempts to develop aerobic bacterial strains for dye decolorization often caused very specific organisms which showed decolorization capability for individual dyes. Additionally, an innovative biological activated carbon (BAC) was developed to allow microorganisms to attach in this study.

2. Model development

The conceptual basis of reactive orange 16 (RO16) from bulk liquid transported into liquid film then diffused into biofilm and activated carbon by intraparticle diffusion is shown in Fig. 1. To model the kinetics of RO16 of adsorption and biodegradation in a completely-mixed reactor, the following assumptions are made [4]: (1) Activated carbon is a homogeneous spherical particle; (2) there is no biological reaction occurred within the activated carbon; (3) A stagnant layer covers the biofilms; (4) no biodegradation occurs inside the activated carbon; (5) local equilibrium occurs at the biofilm- activated carbon interface and (6) the biofilm is homogeneous.

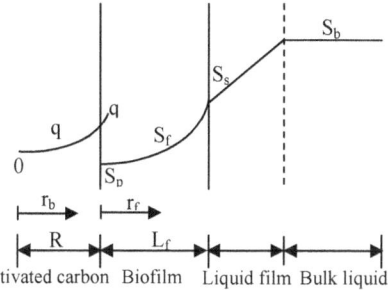

Fig. 1 Concentration profiles for biological activated carbon (BAC)

Two models for intraparticle diffusion are commonly employed: pore diffusion and homogeneous solid diffusion. A homogeneous solid diffusion model, or simply a solid diffusion model, is used in this study. The equation is a typical diffusion equation in spherical coordinates [5]:

$$\frac{\partial q}{\partial t} = \frac{D_b}{r_b^2} \frac{\partial}{\partial r_b} \left(r_b^2 \frac{\partial q}{\partial r_b} \right) \tag{1}$$

where q is the surface concentration of adsorbed RO16 (M_s/M_b); D_b is the surface diffusivity of RO16 (L^2/T); r_b is the radial coordinate in activated carbon (L); and t is time (T).

The RO16 utilization rate by *F. trogii* in the biofilm reactor can be described by the following equation [6]:

$$\frac{\partial S_f}{\partial t} = \frac{D_f}{r_f^2} \frac{d}{dr_f} \left(r_f^2 \frac{dS_f}{dr_f} \right) - \frac{\mu_m}{Y} \frac{S_f}{K_s + S_f} X_f \tag{2}$$

where S_f is the RO16 concentration in the biofilm (M_s/L^3); μ_m is the maximum specific growth rate (1/T); Y is the yield coefficient of *F. trogii* (M_x/M_s); K_s is the half-velocity coefficient of RO16 (M_s/L^3); and X_f is the biofilm density of *F. trogii* (M_x/L^3). The RO16 and *F. trogii* concentrations in the bulk liquid in completely-mixed biofilm reactor can be described as following equations, respectively [6]:

$$\frac{dS_b}{dt} = \frac{Q}{V\varepsilon}(S_{b0} - S_b) - k_f(S_b - S_s)\frac{3X_w(R + L_f)^2}{V\varepsilon\rho_p R} - \frac{kX_bS_b}{K_s + S_b} \tag{3}$$

$$\frac{dX_b}{dt} = \left(\frac{YkS_b}{K_s + S_b} - k_d - \frac{Q}{V\varepsilon} \right) X_b + \frac{3X_w b_s L_f X_f}{V\varepsilon\rho_p R} \tag{4}$$

where S_b is RO16 concentration in the bulk liquid (M_s/L^3); k_d is the decay coefficient (1/T); Q is flow rate (L3/T); V is the effective reactor volume (L3) and ε is the reactor porosity (dimensionless).

3. Materials and methods

The fungal strain *F. trogii* ATCC 200800 is known to be able to degrade various dyes [7]. The fungal strain was grown on potato dextrose agar (PDA) plates at 28°C for 7 days and was stored at 4°C. The experiments were performed not only by cultivation on a solid phase but also by cultivation in liquid. The fungal strain was pre-cultured and it was prepared for the small pieces (the disk size of 1 cm² mycelium) on PDA [8,9]. A laboratory-scale completely-mixed biofilm reactor was setup and conducted using RO16 as a model substrate. The bioreactor system consisted of feed tank, pH and DO controllers, a main body of biological activated carbon (BAC) and sampling port. The pure culture of *F. trogii* ATCC 200800 was mixed with the activated carbon before being put into the completely-mixed biofilm reactor. RO16 was measured using UV-vis spectrophotometer (Shimadzu, model UV-1700) at 568 nm [1]. The uninoculated dye free medium was used as blank. All assays were performed in duplicate.

3.1 Supporting media

The specifications of granular activated carbon for G-340 were described as follows: particle size: 8 x 30 mesh; mean particle diameter: 0.9-1.1 mm; hardness: > 93%; bulk density: 0.46-0.50 g/cm³; total surface area:> 950 m²/g. Therefore, the total surface area was about 1.92×10^5 cm².

3.2 BAC-reactor configuration

The main body of the BAC-reactor with 9 cm in diameter and 132 cm in the bed length is shown in Fig. 2. The BAC-reactor with a high recycle flow rate ($Q_r/Q = 20$) to maintain a completely nixed column reactor was conducted to evaluate the kinetics of RO16 decolorization by *F. trogii*. The feeding stream containing nutrient medium and reactive-dye substrate (10 mg RO16/L) was continuously pumped upward from the lateral side into the column throughout the distributed plate. The organic loading rate of RO16 was 0.08 kg/m³-d. Samples were collected from the effluent at the designated time intervals to measure the residual dye concentration and suspended cells concentration. On the bottom of reactor, a sedimentation zone was designed to accumulate decay and shear-off biomasses throughout the entire experiment. The effective volume of reactor is 8.4 L, which yields a hydraulic retention time (HRT) of 3 h. The reactor temperature was controlled at 28°C using an automatic moisture-proof heater. The pH was maintained at 7.3-8.4 and DO was maintained at 7.8-8.2 in the influent throughout the experimental test.

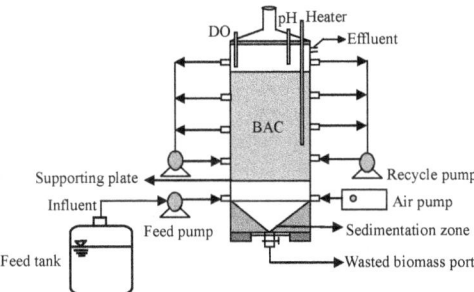

Fig. 2 A laboratory-scale biofilm reactor system.

3.3 Analytical methods

Reactive orange 16 was measured using UV-vis spectrophotometer (Shimadzu, model UV-1700) at 568 nm [1]. The uninoculated dye free medium was used as blank. All assays were performed in duplicate and compared with uninoculated controls. Viable cells were counted by using serial dilution technique and confirmed by plating on nutrient agar and incubated at 28°C for about 24 h to determine *F. trogii* cells concentration.

4. Results and discussion

4.1 Biokinetic and reactor parameters

The input parameters to the model for RO16 decolorization by *F. trogii* in BAC-process were listed in Table 1. The parameters are categorized as follows: (1) measured by kinetic tests; (2) calculated from empirical formula listed in literature; (3) assigned as needed to fit the experimental data.

Table 1 Input parameters to kinetic BAC-model system

Parameter	Symbol	Value	Unit
Measured			
RO16 concentration in the feed	S_{b0}	10	mg RO16/L
Freundlich isotherm coefficient	n	1.755	dimensionless
Freundlich isotherm coefficient	K_p	0.271	$(g/g)(L/mg)^{1/n}$
Maximum specific utilization rate of RO16	k	3.1	mg RO16/mg cell-day
Half-velocity coefficient of RO16	K_s	107.3	mg/L
Yield coefficient of *F. trogii* cells	Y	0.202	mg cell/mg RO16
Decay coefficient of *F. trogii* cells	b	0.024	day^{-1}
Concentration of suspended *F. trogii* cells in the feed	X_{b0}	4.8×10^{-3}	mg/cm^3
Surface diffusivity of RO16	D_s	6×10^{-4}	cm^2/day
Influent flow rate	Q	6.7×10^4	cm^3/day
Effective reactor volume	V	8.4×10^3	cm^3
Total surface area of activated carbon	A	1.92×10^5	cm^2
Reactor porosity	ε	0.54	dimensionless
Radius of activated carbon particle	R	0.05	cm
Apparent activated carbon density	ρ_p	0.48	g/cm^3
Weight of activated carbon	X_w	1.54×10^3	g
Calculated			
Liquid film transfer coefficient	k_f	346.8	cm/day
Diffusion coefficient in the biofilm	D_f	0.192	cm^2/day
Shear-loss coefficient of biofilm	b_s	0.253	day^{-1}
Density of *F. trogii* biofilm	X_f	10.64	mg cell/cm^3
Assigned			
Initial *F. trogii* biofilm thickness	L_{f0}	5.5×10^{-4}	cm

4.2 Reactive orange 16 decolorization

The model-predicted and experimental results for decolorization of RO16 are shown in Fig. 3(a). The RO16 concentrations are all normalized with respect to the influent substrate concentration. The model simulated the experimental results fairly well throughout the entire course of the test. The substrate concentration first increased steadily to about 0.6 mg/L (0.06 S_{b0}) at 0.2 days. At this period of time, there was no significant biological growth inside the reactor and no detectable biodegradation of the RO16 by *F. trogii* in the reactor. The reactor was behaving similar to an activated-carbon-adsorber, and the substrate-concentration curve was the same as a typical breakthrough curve of an activated carbon adsorber. At this stage, activated carbon was adsorbing substrate without significant resistance to the diffusion of the RO16 posed by the *F. trogii* biofilm. The second part of the RO16 curve ran from 0.2 days to 2 days, when the RO16 curve started to deviate from the breakthrough curve of a carbon adsorber. The effluent concentration of RO16 decreased rapidly. Apparently, attached and suspended *F. trogii* cells were actively utilizing the RO16 during this period. Meanwhile, attached and suspended *F. trogii* cells were actively growing at this time. The model was able to predict reactor performance fairly well during this transient period. The third part of the substrate curve ran from 2 days to 4 days. At this period, the experimental data were lower than modeling results for the effluent of substrate. The reason was that the microorganisms grew enough to form biofilm. Thus, the effect of shear loss became insignificant although biofilm became thicker. The lower shear loss results in lower suspended biomass. The suspended cells decomposed and released soluble microbial products (SMP) was insignificant in the effluent concentration of RO16 at this period of time. The fourth part of the substrate curve ran from 4 days to 7.9 days. At this period, the BAC-process reached a steady-state condition and the effluent of substrate was about 0.2 mg/L (0.02 S_{b0}). The removal efficiency for substrate was about 98% at this period. As can be seen, the model prediction is in fair agreement with the experimental result. Although the soluble microbial products (SMP) would be released by biofilm at this stage, the SMP produced by the biodegradation of RO16 would be adsorbed by activated carbon, which made SMP not a significant interference to the effluent.

4.3 Growth of suspended cells

Fig. 3(b) shows the generating suspended biomass varied with time. The growth concentration curve went through log-growth phase then reached a steady-state condition. The elapsed time required for suspended *F. trogii* cells to reach a steady-state condition is almost the same with that for substrate biodegradation. This indicated that the suspended *F. trogii* cells reached a maximum growth rate while substrate had a maximum utilization rate at a steady-state condition. The maximum growth at steady-state condition was maintained at 18.6 mg cell/L.

4.4 Flux into biofilm and activated carbon

There are two fluxes of RO16 in the BAC-process; the flux from the liquid phase into the *F. trogii* biofilm (J_b) and the flux from the *F. trogii* biofilm into the activated carbon (J_p). J_b and J_p varied with time in the BAC-reactor are plotted in Fig. 3(c). The figure shows that J_b and J_p were equal at the beginning of the test because the utilization of *F. trogii* biofilm was negligible at this time. Most of the RO16 was decolorized by activated carbon adsorption. The two curves started to deviate with one another around 0.2 days, when the *F. trogii* biofilm started to grow actively. The J_b value for BAC-model started out at a value of 0.08 mg/cm^2-day that was controlled by adsorption when the utilization of *F. trogii* biofilm was zero initially.

J_p in the BAC reactor decreased first due to the fast utilization of the RO16 by the *F. trogii* biofilm, which was beginning to develop. As the *F. trogii* biofilm accumulated, J_p decreased to zero and then went negative, which signaled bioregeneration. The flux J_p changed from a positive value to a negative value at 0.8 days. This required a reversal of the adsorbate density gradient inside the activated carbon. The resulting negative adsorbate density gradient caused the substrate to diffuse out the activated carbon because of the growth of biofilm. As the biofilm grew thicker, the biofilm consumed RO16; thus, the concentration of RO16 at the biofilm/activated carbon interface became lower. As bulk and biofilm concentration declined, they became lower than the concentration at equilibrium with the previously adsorbed substrate. At this point, RO16 simply diffused out of the activated carbon and activated carbon was bioregenerated by the *F. trogii* biofilm. As the steady-state approached, J_p went to 0 asymptotically from a negative value, while J_b approached a constant value. The curves showed that *F. trogii* biofilm utilization became the dominant mechanism responsible for the RO16 removal at steady-state in the BAC-reactor.

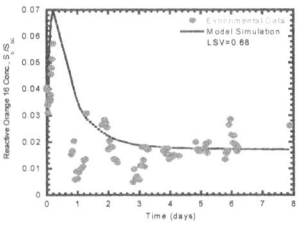 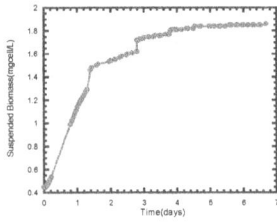 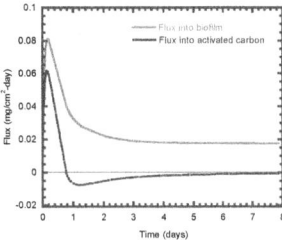

Fig. 3 Experimental data and model simulation (a) RO16 effluent concentration (b) suspended *F. trogii* cells concentration in effluent and (c) Flux into biofilm and activated carbon

5. Conclusions

This study demonstrates the feasibility of using *F. trogii* biofilm and suspended *F. trogii* cells for continuous decolorization of dye-laden influents using BAC-process. The kinetic BAC-model and experimental approaches elucidated the total mechanisms governing the interactions of the activated carbon adsorption and biofilm biodegradation for reactive-dye decolorization by *F. trogii* cells. The BAC-model was able to predict performance fairly well for a non-steady-state experiment. The *F. trogii* Biofilm bioregenerated the activated carbon by lowering the RO16 concentration at the biofilm/activated carbon interface. The RO16 previously adsorbed inside the activated carbon simply desorbed out of the activated carbon through a reversal of flux entering activated carbon. The approaches of experiments and kinetic model presented in this paper can be employed for the design of a pilot-scale or full-scale BAC-process for reactive-dye decolorization by *F. trogii* cells in textile wastewater.

Acknowledgements This research was supported in part by a grant from the Ministry of Science and Technology of Taiwan under Contract No. NSC 100-2221-E-166-004-MY2.

References

[1] Park C, Lee M, Lee B, Kim SW, Chase HA, Lee J, Kim S. Biodegradation and biosorption for decolorization of synthetic dyes by *Funalia Trogii*. Biochemical Engineering Journal. 2007; 36:59-65.
[2] Choi HD, Shin MC, Kim DH. Removal characteristics of reactive black 5 using surfactant-modified activated carbon. Desalination. 2008; 223:290-98.
[3] Rao KR, Srinivasan T, Venkateswarlu Ch. Mathematical and kinetic modeling of biofilm reactor based on ant colony optimization. Process Biochemistry. 2010; 45:961-72.
[4] Liang CH, Chiang PC, Chang EE. Modeling the behaviors of adsorption and biodegradation in biological activated carbon filters. Water Research. 2007; 41:3214-50.
[5] Smith E, Ghiassi K. Chromate removal by an iron sorbent: mechanism and modeling. Water Environment Resesrch. 2006; 78:84-93.
[6] Tsai HH, Ravindran V, Pirbazari M. Model for predicting the performance of membrane bioadsorber reactor process in water treatment applications. Chemical Engineering Science 2005; 69:5620-36.
[7] Arora DS, Chander, M, Gill PK. Involvement of lignin peroxide, manganese peroxide and laccase in degradation and selective of wheat straw. International Biodeterioration & Biodegradation. 2002; 50:115-20.
[8] Kapdan IK, Kargi F, McMullan G, Marchant R. Effect of environment conditions on biological decolorization of textile dyestuff by *C. versicolor*. Enzyme and Microbial Technology. 2000; 26:381-7.
[9] Jang MY, Ryul WR, Cho MH. Laccase production from repeated batch cultures using free myxerlia of *Trametes* sp. Enzyme and Microbial Technology. 2002; 30:741-6.

Antibacterial activity of an aromatic plant from Algerian kitchen

A. Attou[*,1]**, A. Benmansour**[1]**, F. Haddouchi**[1] **and I. Amrani**[2]

[1]Laboratory: Natural Products: biological activities and synthesis (LAPRONA). Faculty of Sciences of Nature and Life & Sciences of Earth and Universe, Department of Biology. University Abou Bekr Belkaid, B.P.119 Tlemcen, 13000, Algeria.
[2] Laboratory ToxiMed, Faculty of Sciences of Nature and life & Sciences of earth and Universe, University Abou Bekr Belkaid, B.P.119 Tlemcen, 13000, Algeria.
*Corresponding author: e-mail: heyfraise21@yahoo.fr, Phone: +213552562910

Ruta chalepensis L. is an aromatic plant belonging to the family of *Rutaceae*, commonly called by locals "Fidjel". It is spontaneous, largely spread in North Africa, especially Algeria.
The rue is a medicinal aromatic plant still used in traditional medicine in many countries as a laxative, anti-inflammatory, analgesic, antispasmodic, abortifacient, antiepileptic, emmenagogue and for the treatment of skin diseases, and very used in many dishes from the aromatic oil and cheese to flavoring any kind of prey.
The extraction by hydrodistillation of the essential oil of the whole aerial plant has significant returns up to 1,90%. The essential oil has a medium antibacterial activity against staphylococci and streptococci, and no activity or very low against enterobacteriaceae species.

Keywords: bioactive molecules; essential oil; antimicrobial activity; *Ruta chalepensis.*

1. Introduction

Since time immemorial, Men appreciate the soothing and analgesic virtues of plants.

The active principles of medicinal plants are often related to the products of secondary metabolism. Their properties are currently recognized and for a good number listed, and thus harnessed, in the context of traditional medicines and also in modern allopathic medicine [1, 2, 3].

The family *Rutaceae* consists of 150 genera and 1500 species of treelets known to accumulate essential oils, flavonoids, coumarins, and several sorts of alkaloids.

Ruta chalepensis (fig n°1), herbaceous plant with woody stem at the base, can reach 1 m in height, with aromatic foliage, and dark yellow flowers, with four or five petals; it is an ornamental plant in gardens.

The name *Ruta* comes from the greek « rhyté » which means rescued, prevent, or "REO" means flowing certainly making reference to its emmenagogue properties [2,4], and commonly known by the RUE.

In cooking, it is used to flavor sauces and prey, cheese, oil and vinegar [4, 5, 6],

It was used as emmenagogue and abortifacient, it has a clear stimulating effect on the uterus, until today it is so used as : Analgesic; Anti-inflammatory; antiseptic; antispasmodic; aphrodisiac; cardiotonic; sedative; stomachic; the aqueous extracts of the plant has a hypotensive activity by a direct effect on the cardiovascular system [7, 8].

2. Materials and methods

2.1. Plant material

The aerial parts of the plant were collected at flowering season from a station in west Algeria, and then dried at room temperature for two weeks.

Fig n°1: *Ruta chalepensis*

2.2. The extraction of essential oil

It was done by steam distillation of the aerial part of the plant, where 100 g of dry plant is introduced into a flask bi collar, and moistened with water; the mixture is brought to a boil for 2-3 hours. The water laden vapours of essential oil, the refrigerant passing through, condense and drop into a separator funnel; water and oil separate by density difference [1, 9]. The essential oil is stored in dark vials until use.

2.3. Evaluation of the antimicrobial activity

NCCLS, 1990 [10], the activity of essential oils of *Ruta chalepensis* was evaluated by the method of discs on agar, which is a qualitative technique based on measuring the diameters of inhibition in mm.

The antimicrobial activity was evaluated against a sample of wild strains (sample of 131 strains: 101enterobacteriaceae, 14 staphylococci and 16 streptococci) from the Laboratory of Microbiology at the Regional Military University Hospital of Oran during 2 months from 05/01/2014 to 06/03/2014.

The culture of the strains is measured par turbidity; and adjusted to 0.5 standard of McFarland scale, corresponding to 1-2 x 10^8 CFU/ml.

100µl of the suspension used to inoculate the agar plates, then filter paper discs impregnated with the essential oil are disposed, the Petri dishes are incubated at 37°C for 24h (fig n°2).

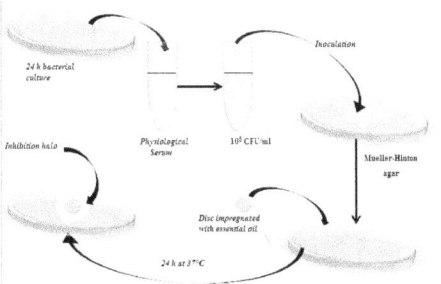

Fig n°2: Evaluation of the antimicrobial activity

3. Results

3.1 Essential oil isolation

The dried aerial parts of the plant subjected to a hydrodistillation have revealed a yield of about 1.90±0.01% (w/w).

3.2 Antimicrobial activity

Table 1 shown the diameters of inhibition of the essential oil against the bacteria tested. The results indicate that the essential oil of *Ruta chalepensis* has low antibacterial activity against all strains; with inhibition zones don't exceed the 15 mm, and no activity against the enterobacterial species.

Table 1 In vitro antibacterial activity of *Ruta*'s essential oil.

Strains	Sample number	Inhibition diameter (mm)		
		6	7-10	10-15
Pseudomonas aeruginosa	13	13	0	0
Escherichia coli	49	47	2	0
Enterobacter cloacae	5	5	0	0
Proteus mirabilis	11	11	0	0
Klebsiella pneumoniae	14	14	0	0
Acinetobacter baumanii	9	9	0	0
Staphylococcus aureus	14	5	3	6
Streptococcus sp.	16	4	5	7

4. Discussion

The yields of the essential oil found in our samples are higher than those recorded by Merghache and *al*, 2009 (0.82 %) [9], but lower than those recorded by Mejri and *al*, 2010 (up to 5.51 % v / m) on the same species of northern Tunisia [11].

Previous researches confirmed that the chemical composition of the essential oil refers on his capacity of antibacterial activity. These studies (Merghache and al, 2009[9]; Ben-Bnina and *al*, 2010[12]; Mejri and *al*, 2010 [10]; Haddouchi and *al*, 2013[13]) found that the essential oils of *Ruta* species are characterised by the presence of more than 81% of ketones, This composition confirmed the less potential of antibacterial activity, but the important antifungal activity proved by the previous studies, that's why this plant is still used to the preservation and flavoring of prey cheeses and vinegars.

5. Conclusion

Our study belongs to the programme of the evaluation of the medicinal plants growing in Algeria, and comes to confirm some results from previous works:
-   The essential oils from *Ruta* species have low potential of activity against bacteria, but with the use of contact method of evaluation;
-   The yields of essential oil from our plant show that the climate can affect the composition and performance of plant essential oil;
-   The composition of this essential oil and the other secondary metabolites of *Ruta chalepensis* can open more possibilities to laboratory assay to search for others activities.

References

[1] Schiller C. and Schiller D.; 1994; 500 Formulas for Aromatherapy (Mixing Essential Oils for Every Use); Ed: STERLING PUBLISHING; p: 11-22.
[2] Iserin P. and *al*; 2001; Encyclopaedia of medicinal plants; Ed 2: LAROUSSE; p: 1-14.
[3] Doerper S.; 2008; Modification of Furocoumarine's synthesis in *Ruta graveolens L.* by metabolic engineering approach; University Nancy, INRA; p: 12 - 34.
[4] Baba Iassa F.; 1999; Useful Plant Encyclopedia: Flora of Algeria and the Maghreb; Ed: MODERN LIBRARY ROUIBA; p: 243 - 244.
[5] Takhtajan A.; 2009; Flowering Plants; Ed 2: SPRINGER; p: 33 - 41, 375.
[6] Bonnier G.; 1999; The biggest flora in colored edition (La Grande Flore en Couleur); Ed: BELIN; Tome 3; p: 205 - 206.
[7] Wiart C.; 2006; Medicinal Plants of the Asia – Pacific: Drugs for the future? ; Ed: WORLD SCIENTIFIC; p: 401 - 416.
[8] Foster S. and Tyler V.E.; 1999; A Sensible Guide to the Use of Herbs and Related Remedies; Ed 4: Tyler's honest herbal, HAWORTH HERBAL PRESS; p: 325-326.
[9] Merghache S., Hamza M. and Tabti B.; 2009; Physicochemical study of the essential oil of *Ruta chalepensis L.* From Tlemcen, Algeria; Africa Science 5 (1); p: 67- 81.
[10] National Committee for Clinical Laboratory Standards; 1990; Performance standards for antimicrobial disk susceptibility tests; NCCLS; Villanova.
[11] Mejri J., Abderrabba M. and Mejri M.; 2010; Chemical composition of the essential oil of *Ruta chalepensis L.*: Influence of drying, hydrodistillation duration and plant parts; Industrial Crops and Products 32, Ed: ELSEVIER; p: 671- 673.
[12] Ben-Bnina E., Hammami S., Daamii-Remadi M., Ben-Jannet H. and Mighiri Z.; 2010; Chemical composition and antimicrobial effects of Tunisian *Ruta chalepensis L.* essential oils; Journal de la Société Chimique de Tunisie; 12; p:1-9
[13] Haddouchi F., Chaouche M.T., Yosr Z., Riadh K., Attou A. and Benmansour A.; 2013; Chemical composition and antimicrobial activity of the essential oils from four Ruta species growing in Algeria; Food Chemistry; 141. Ed: Elsevier; p: 253-258

Antibacterial activity of multiple plant essential oils and their potential use as food preservatives

J. Thielmann[*] **and C. Hauser**

Fraunhofer Institute for Process Engineering and Packaging IVV, Giggenhauser Straße 35, 85354 Freising, Germany
*Corresponding author: e-mail: julian.thielmann@ivv.fraunhofer.de, Phone: +49 8161 491 600

It is widely understood that plant essential oils (EO) exhibit inhibitory potential against food borne pathogenic bacteria, which has put them into focus of intensive research regarding food preservation and safety. Nonetheless there is still very fragmentary information on the antimicrobial potency of essential oils from various plant genera. Due to the variety of EO containing plants, the use of inconsistent methods for antimicrobial susceptibility testing and the fact that information on non-antimicrobial EOs often remains unpublished, it is challenging to adjust penetrative microbiological research on less investigated EOs. Therefore it is necessary to extend data on antimicrobial activity of plant EOs considering standardized microbiological examination as well as chemical characterization. This study should provide additional data in means of minimal inhibitory concentrations (MIC) of a multitude of plant EOs against food borne pathogen bacteria (i.e. *Escherichia coli, Staphylococcus aureus*) acquired by an accepted microdilution assay for antimicrobial susceptibility testing (CLSI standard), without the influence of organic solvents or emulsifying agents. Building a consistent database, aggregating information on antibacterial active and non-active plant essential oils, will help to identify the most promising EOs regarding further investigations on their use as plant-derived natural food preservatives.

Keywords antimicrobial activity; essential oils; foodborne pathogenic bacteria; food safety; food preservation

1. Introduction

The antimicrobial potential of essential oils (EO) has attracted the attention of scientists for decades. Their traditional use in folk medicine applications distinguishes them as potential sources of powerful natural antimicrobial compounds. Furthermore they are often reckoned to be promising alternative agents for food preservation [1]. The recent changes in modern consumer behaviour, turning from convenient processed foods towards "green" convenience products, has initiated the search for plant extracts which possess significant antimicrobial activity to act as natural preservatives, ensuring microbial shelf life and food safety without alteration of the overall green character of certain food products. In this context many results have been published regarding the antimicrobial activities of essential oils from various plant species, but overall information remains fragmentary [2]. The wide variety of factors influencing the chemical composition of EOs and the inconsistent use of strongly differing methods aggravate the comparison and assessment of the antimicrobial activities of essential oils. In addition many results appear to remain unpublished due to unsatisfactory activity. In this study we compared the reliability of two methodological approaches based on a standardized microdilution assay for susceptibility testing [3]. *Litsea cubeba* essential oils and its major compound citral were used to identify the more reliable method, which was subsequently used to screen a wide variety of essential oils for their antibacterial activity against the food borne pathogenic bacteria *Escherichia coli* and *Staphylococcus aureus*. The screening comprised a wide variety of different essential oils. As far as available EOs from each plant species were purchased from three different manufacturers, to compare the activity of such standardized EO products. The results should help to identify further promising candidates which will be investigated towards their potential use as crude biobased food preservative agents.

2. Materials and Methods

2.1 Bacterial strains, growth conditions and preparation of inoculums

The pathogenic bacteria *Escherichia coli* DSM 1103 (Gram-negative) and *Staphylococcus aureus* DSM 1104 (Gram-positive) were purchased from the Leibniz Institute DSMZ - German Collection of Microorganisms and Cell Cultures (DSMZ, Braunschweig, Germany). The strains are recommended for antimicrobial susceptibility testing and were used as surrogates for foodborne strains of the respective species. Cells were cultured initially in sterile tryptone soy broth (TSB) at 37 °C in a shaking bath. Populations were harvested after 18 h from the late log-phase and washed twice by centrifugation at 9000 g for 10 min. Supernatant was substituted by sterile

¼*Ringer's solution. Cell count was adjusted in ¼*Ringer's solution to $1.0*10^8$ cfu/mL by turbidity measurement at 600 nm using a 0.5 McFarland standard. Suspension was used within 15 minutes.

2.2 Essential oils

Each oil variety was purchased from at least three different German manufacturers, as far as available. The oils were declared as naturally pure by the manufacturers and were free of solvents or additives. The resealable vials were stored at 5 °C in the dark and allowed to adjust to room temperature prior to investigation.

2.3 Broth microdilution assay

The antibacterial activity of each EO was assessed in a broth microdilution assay, following CLSI guidelines for antimicrobial susceptibility testing [3] with slight optimizations regarding EO incorporation (see 2.3.1 and 2.3.2). Assays were performed in sterile 96 well microplates (Greiner, Frickenhausen, Germany). EO concentrations tested ranged in bisecting dilution steps from 6400 to 50 µg/mL. Growth, sterility and quality controls were performed for each trial. Chloramphenicol was used as antibiotic standard for quality controls. Mueller-Hinton broth (MHB) (Merck, Darmstadt, Germany) was used as appropriate growth medium. Sterile deionized water served as dilution medium. All media were adjusted to pH 7.0. Inoculum was prepared by a hundred fold dilution of the adjusted cell suspension in MHB. Final cell density averaged out $5.0*10^5$ cfu/well. Each EO was tested in triplicate per plate. A fourth column served as blanks, to include possible turbidity from the EOs into OD calculations. Absorption measurement was performed after 24 h incubation at 37 °C in a microplate reader (Tecan, Männedorf, Switzerland) at 595 nm. Plates were incubated with and measured without the lid. Each well was measured at nine spots with 5 flashes per spot. Each plate was preliminary shaken orbital for 1 min. The minimal inhibitory concentration (MIC) was defined as the lowest concentration which did just not result in turbidity by cell growth within 24 h.

2.3.1 Dispersion

The dispersion method, as foremost published by Remmal et al. (1996) [4], is based on growth and dilution media with increased viscosity, to slow phase separation between EO droplets and medium matrix. Each medium contained 0.20 % of the polysaccharide agarose (Merck, Darmstadt, Germany) which was added prior to sterilization [4]. All dilution steps were performed using deionized water containing 0.20 % agarose. To sufficiently disperse the EO into the dilution medium, screw capped tubes were used, which were shaken vigorously for 30 s. Afterwards the tubes were vortexed slightly to drive out air bubbles. The dispersions remained stable for 48 h.

2.3.2 Emulsion

Growth and dilution medium were spiked with 0.20 % of Tween 80 (Merck, Darmstadt, Germany) prior to sterilization. Tween 80 served as an emulsifying agent to improve EO distribution in the media. In this approach each EO was furthermore gathered in 1 g sterile Dimethylsulfoxide (DMSO) (Sigma-Aldrich, St. Louis, MO, US) prior to a tenfold dilution in the dilution medium containing Tween 80. The final DMSO concentration in the well was 5 %. To sufficiently distribute the EO into the dilution medium, screw capped tubes were used, which were shaken vigorously for 30 s. Afterwards the tubes were vortexed slightly to drive out air bubbles. The emulsions remained stable for 30 h. DMSO and Tween 80 did not affect cell growth.

3. Results and Discussion

3.1 Dispersion vs. emulsion approach

A crucial factor regarding susceptibility testing of essential oils is the applied approach of oil incorporation into the test media. The use of organic solvents in combination with emulsifying agents, appear to be the most common approach [5]. Other studies follow the findings of Remmal et al. (1993) who stated, that ethanol and tween 80 reduce the inhibitory activity of EOs against microorganisms. Regarding the wide screening of different EOs we preliminary compared both approaches based on the broth microdilution assay following CLSI guidelines. The results for *Litsea cubeba* essential oils from three different manufacturers against *E. coli* and *Staph. aureus* are given in Table 1 in terms of the respective MIC values. The data shows that lower MIC values can be achieved by using the dispersion approach. Emulsified EOs revealed higher MICs against both bacterial species in almost every case. Results for emulsified EOs versus *Staph. aureus* appear to be the most inconsistent. The explanation for these deviations may be found in the microstructural differences of emulsions and dispersions, such as micelle size or distribution homogeneity [4], but have not been further investigated to date. The results highlight the necessity of a standardized method for antimicrobial susceptibility testing of microorganisms against essential oils. A reliable comparison of MICs as degree of antimicrobial activity of crude essential oils is already strongly aggravated when the values are assessed by methods with only slight variations [1, 2].

Table 1 Minimal inhibitory concentrations (MIC) of three different *Litsea cubeba* essential oils versus *E. coli* DSM 1103 and *Staph. aureus* DSM 1104 tested in a broth microdilution assay based on a dispersion or an emulsion approach [13].

Bacterial species	Method	*L. cubeba* type	MIC [µg/mL]
E. coli DSM 1103	Dispersion	1	200
		2	200
		3	200
	Emulsion	1	800
		2	800
		3	800
Staph. aureus DSM 1104	Dispersion	1	100
		2	100
		3	100
	Emulsion	1	400
		2	200
		3	100

The same comparative trials were performed using pure citral (Sigma-Aldrich, St. Louis, MO, US), which is a mixture of the cis-trans-isomeric terpenoids geranial and neral. Citral is the major compound in *L. cubeba* essential oils and known to possess considerable antibacterial activity alone [6]. Corresponding results are presented in Table 2. The given MICs differ in dependence on the respective incorporation approach used. But in contrast to the results for the crude EOs of *L. cubeba* pure citral achieved lower MICs in the emulsion approach than in the dispersion variant. This may be attributed to the complexity of the crude essential oil and better distribution of the single compound citral in the growth medium. Regardless of the incorporation approach used, citral did not have higher activity as the crude *L. cubeba* essential oils against *E. coli*. But versus *Staph. aureus* the MIC for citral was 50 µg/mL which is half of the lowest MIC of the crude extracts (100 µg/mL *L. cubeba* type 3). The antibacterial mechanism of action of citral is only partially revealed to date, but may be mainly due to cell envelope damages [7]. Deviations in MIC values between *E. coli* and *Staph. aureus* may therefore be attributed to structural differences between Gram-negative and Gram-positive bacteria.

Table 2 Minimal inhibitory concentrations (MIC) of pure citral versus *E. coli* DSM 1103 and *Staph. aureus* DSM 1104 tested in a broth microdilution assay based on a dispersion or an emulsion approach [13].

Bacterial species	Method	MIC [µg/mL]
E. coli DSM 1103	Dispersion	400
	Emulsion	200
Staph. aureus DSM 1104	Dispersion	200
	Emulsion	50

3.2 Screening results

A wide variety of essential oils from multiple plant species was tested for their antibacterial activity against *E. coli* and S*taph. aureus* in a dispersion based broth microdilution assay. The overall results can only be presented in a condensed manner. Table 3 gives an overview of selected results of those EOs which were most active and partially available from three different distributors. The data highlight the strong MIC variations between single oils from different sources. These differences in MIC values are certainly due to variations in the chemical composition of the respective EOs. Although these findings are mostly exemplarily they show the necessity of chemical analyses of the tested oils to allow reliable result comparisons and conclusions on activity on a molecular level. The data clearly illustrates the elevated activity of most EOs versus the Gram-positive bacterium *Staph. aureus*. Besides certain exceptions most EOs did not show lower MIC values than 200 µg/mL. Regarding future applications of essential oils as bioderived food preservative agents the screening was mainly performed to identify the EOs with the highest antibacterial potential. Compared to in vitro studies the application of antimicrobial compounds in food products is often accompanied with a strong loss of activity [8]. Therefore it was defined that only EOs with MICs lower than 200 µg/mL should be considered as effective.

Table 3 Minimal inhibitory concentrations (MIC) of the most effective essential oils versus *Escherichia coli* DSM 1103 and *Staphylococcus aureus* DSM 1104 tested in a broth microdilution assay based on a dispersion (n. i. = no inhibition; n. t. = not tested to date) [13].

| Essential oil | | MIC [µg/mL] | | | | | |
| | | *E. coli* DSM 1103 | | | *Staph. aureus* DSM 1104 | | |
Common name	Plant name	1	2	3	1	2	3
Citronella	*Cymbopogon nardus*	200	n. a.	n. i.	200	n. a.	n. i.
Chinese cinnamon	*Cinnamomum cassia*	200	n. a.	50	200	n. a.	50
Lemongras	*Cymbopogon citratus*	400	400	400	400	100	400
Litsea	*Litsea cubeba*	200	200	200	50	100	100
Niaouli	*Melaleuca quinquenervia*	6400	n. I.	3200	200	100	1600
Oregano	*Origanum vulgare*	50	100	50	50	50	50
Palmrose	*Cymbopogon martinii*	200	200	400	400	200	800
Patchouli	*Pogostemon cablin*	n. i.	n. i.	n. i.	50	200	200
Citronella	*Cymbopogon nardus*	200	n. a.	n. i.	200	n. a.	n. i.

The antimicrobial activity of essential oils is highly dependent on their chemical composition. This composition is influenced by a set of factors regarding plant growth such as origin, environmental situation or point of harvest [2]. A single comparison of the antimicrobial activity of plant EOs by the plant species name is therefore insufficient. This study has shown that even oils which are standardized by manufacturer's demands differ greatly in their antibacterial activity. But regarding their application in food matrices as preservative agents it will become necessary to use standardized EOs which are technologically optimized regarding their most active compounds. In view of antimicrobial activities, synergistic effects between contained compounds should be studied in detail as well, to allow further optimizations of applicable EO's activities. These approaches have for example partially been realized for thyme or oregano essential oils and their compound thymol and carvacrol [9, 10] which appear to be the best investigated antimicrobial essential oils and EO compounds. Furthermore there have been many studies regarding their application on food products but sensorial affections of the products have been the major drawbacks [11, 12].

Acknowledgements   Parts of this research was carried out under the auspices of the German Federation of Industrial Research Associations (AiF). It is financed within the budget of the German Federal Ministry of Economic Affairs and Energy (BMWi) through the program to promote collective industrial research (IGF). This project was implemented under the CORNET initiative (IGF-Nr. 9615 EN).

References

[1]   Negi PS. Plant extracts for the control of bacterial growth: Efficacy, stability and safety issues for food application. *International Journal of Food Microbiology.* 2012; 156:1:7–17.

[2]   Burt S. Essential oils: their antibacterial properties and potential applications in foods—a review. *International Journal of Food Microbiology.* 2004; 94: 3: 223–253.

[3]   CLSI - Clinical and Laboratory Standard Institute. Methods for Dilution Antimicrobial Susceptibility Tests for Bacteria That Grow Aerobically; *Approved Standard - Seventh Edition M7-A7.* 2007

[4]   Remmal A, Bouchikhi T, Tantaoui-Elaraki A, Ettayebi M. Inhibition of antibacterial activity of essential oils by tween 80 and ethanol in liquid medium. *Journal de pharmacie de Belgique.* 1993; 48:5:352–356.

[5]   Hammer KA, Carson CF, Riley TV. Antimicrobial activity of essential oils and other plant extracts. *Journal of Applied Microbiology.* 1999; 86: 6: 985–990.

[6]   Chueca B, Pagán R, García-Gonzalo D. Oxygenated monoterpenes citral and carvacrol cause oxidative damage in Escherichia coli without the involvement of tricarboxylic acid cycle and Fenton reaction. *International Journal of Food Microbiology.* 2014; 189:126–131.

[7]   Somolinos M, Garcia D, Condon S, Mackey B, and Pagan R. Inactivation of Escherichia coli by citral. *Journal of Applied Microbiology.* 2009; 108:1928-1939

[8]   Smid EJ, Gorris LGM. Natural antimicrobials for food preservation. In: Rahman MS, editor. Handbook of Food Preservation. New York: Marcel Dekker; 1999. p. 285–308.

[9]   Burt SA, Reinders RD. Antibacterial activity of selected plant essential oils against Escherichia coli O157:H7. *Letters in Applied Microbiology.* 2003; 36:3:162–167.

[10] Didry N, Dubreuil L, and Pinkas M. Antibacterial activity of thymol, carvacrol and cinnamaldehyde alone or in combination. *Die Pharmazie*. 1993; 48:4:301–304.

[11] Emiroğlu ZK, Yemiş GP, Coşkun BK, Candoğan K. Antimicrobial activity of soy edible films incorporated with thyme and oregano essential oils on fresh ground beef patties. *Meat Science*. 2010; 86:2:283–288.

[12] Han JH, Patel D, Kim JE, Min SC. Retardation of Listeria Monocytogenes Growth in Mozzarella Cheese Using Antimicrobial Sachets Containing Rosemary Oil and Thyme Oil. *Journal of Food Science*. 2014; Article first published online: 21 SEP 2014

[13] Fraunhofer Institue for Process Engineering and Packaging IVV, Freising, Germany. internal results. 2014

Antifungal activities of yeasts from oleic environments

J.J. Mateo*, J. Lilao and S. Maicas

Departament de Microbiologia i Ecologia, Universitat de València, C/Dr. Moliner, 50, 46100 Burjassot, Spain
*Corresponding author: e-mail: Jose. J.Mateo@uv.es, Phone: +34 963543008

The traditional olive oil mills separate the decomposed pulp produced at a first stage in a two-phase centrifuge into oil and a liquid solid mixture traditionally called *alpeorujo*. The final solid waste represents about 800 kg per ton of processed olives. This *alpeorujo* still contains 3% residual oil and about 58% water. It can have an important role as a fertilizer or a food additive, provided that it can be detoxified through bioremediation by breaking down the toxic phenolic compounds. Some previous studies have been conducted to evaluate the bacteria and yeast strains involved in this process. In this study we have selected different yeast isolates from different samples dilution, after purification by triple groove, and stored at 4°C for further use. The use of molecular techniques as RFLP analysis and DNA sequencing enabled us to group and identify yeasts isolated from *alperujo*. We have firstly described the presence of *C. norvergica* and *C. molendinolei* and *C. adriatica* (both recently described) on this substrate.

The total yeast strains isolated from the three oil mills were screened for enzymes relevant in the processing of olives, oils and other food products, regarding their potential use in manufacturing to enhance the quality of products. The characterization of some isolates producing β-glucosidase, provides us with a battery of yeasts capable of attack phenolic compounds. This can be used to reduce olive oil bitterness.

The twelve representative yeast isolates, previously selected in the enzymatic characterization were used in a nutritional competition assay. This test was proposed in order to select yeast strains able to affect the co-inoculated fungi when colonizing a common ecological niche. A different inhibitory activity against *Aspergillus flavus* and *A. parasiticus* was found among yeast isolates. *Saccharomyces paradoxus*, *C. boidinii* and *C. oleophila* showed the highest inhibition and *C. molendinolei* and *Candida* sp. isolate the lowest one. Assayed yeast inhibitory activity was affected by the concentration of fungal inoculum: the growth in the lowest fungal inoculum concentration (10^3 spores/mL) was the most inhibited by antagonistic yeasts; similar growth was observed in growth when 10^5 spores/mL were used as inoculum.

Keywords yeasts; olive oil; antifungal

1. Introduction

The traditional olive oil mills separate the decomposed pulp produced at a first stage in a two-phase centrifuge into oil and a liquid solid mixture traditionally called *alpeorujo*. The final solid waste represents about 800 kg per ton of processed olives. This *alpeorujo* still contains 3% residual oil and about 58% water [1]. It can have an important role as a fertilizer or a food additive, provided that it can be detoxified through bioremediation by breaking down the toxic phenolic compounds. Some previous studies have been conducted to evaluate the bacteria [2] and yeast [1] strains involved in this process. Non-*Saccharomyces* yeasts are part of the microbial community contributing to the final organoleptic characteristics of the olive oil production [3-5]. The final quality of olive oil regarding their flavor and other organoleptic attributes is enhanced during the olive fermentation process [6]. The characterization of yeasts involved in this process is nowadays based on molecular techniques, following sequencing on rDNA and PCR-RFLP procedures in order to determine interspecific differences [5, 7]. Moreover, phenotypic assays can contribute to determine the ability of new isolates to conduct the process. Postharvest storage conditions are also considered, to avoid the production of olive oil with a high risk of contamination by mycotoxins. It is known that the olive cake resulting from such olives could present a danger for animals because of the preferential concentration of mycotoxins in oil cakes [8]. Several studies showed the presence of toxigenic moulds, mainly *Aspergillus*, in olives [9, 10]. The scientific community is evaluating the use of living agents to control pests or plant pathogens. This approach is being considered as a reliable alternative to the use of chemical pesticides in field and in post-harvest treatments. The biological approach can contribute to reduce the chemical residues in the olive oil production process. Competition among microorganisms for essential environmental factors, such as nutrients and space, is expected to have a crucial effect on the mycotoxin production of spoilage moulds. Yeasts can be produced in fermenters on cheap media and are able to survive in a wide range of environmental conditions [11].

2. Materials and Methods

2.1 Isolation of yeasts and culture conditions

Yeast strains were isolated from three olive mills located in Eastern Spain (VL, Villamalea; VM, Venta del Moro and O, Ontinyent) during the 2011 and 2012 olive oil production campaigns. The yeasts were isolated as follows: samples (5g) were placed in sterile flasks, mixed with 95 ml sterile saline solution and shaken on a laboratory orbital shaker at maximum speed (500 rev min^{-1}) for 5 hours (adaptation of Katsifas et al. [12]). Mixtures were allowed to settle before making serial dilutions of the supernatant liquid and plating on malt agar extract (20 g/L glucose, 20 g/L malt extract, 2 g/L peptone, 20 g/L agar), tributiryn agar (3 g/L yeast extract, 5 g/L peptone, 10 ml/l tributyrin, 15 g/L agar) and oleic agar (10 g/L peptone, 150 mg/L, CaCl$_2$, 20 g/L agar and 30 ml/l olive oil added after autoclaving). Plates were incubated at 30°C for 72 h. In order to promote growth of strains occurring in low frequencies, 100 mL-flasks containing enrichment media (1 g/L K$_2$HPO$_4$, 0.6 g/L KH$_2$PO$_4$, 1 g/L (NH$_4$)$_2$SO$_4$, 0.2 g/L MgSO$_4$·7 H$_2$O, 0.1 g/L CaCl$_2$, 0.05 g/L FeCl$_3$, 0.3 g/L NaCl$_2$, 0.5 g/L yeast extract, 0.2 g/L glucose and 30 ml/L olive oil added after autoclaving) were inoculated with 5 g of samples and incubated at 30°C for 36 h. Serial dilutions were also plated on the media above described to purify yeast colonies, which were then transferred to malt agar and stored at 4°C. After visual inspection, yeasts showing different morphologies were purified and stored at -20°C in YPD, until further assays. Chemicals and culture media compounds were purchased from Sigma (St. Louis, USA) and Pronadisa (Madrid, Spain) respectively.

2.2 Molecular characterization

Pure cultures of yeasts were grown on 5 mL of YPD medium (1% yeast extract, 2% mycological peptone, 2% glucose) at 28°C for 24-48 h on an orbital shaker. Small-scale preparation of chromosomal DNA was performed by using Ultraclean Microbial DNA isolation Kit (MoBio, Carlsbad, USA). The quality of the DNA was checked by nucleic acid electrophoresis as described elsewhere. DNA was stored at -20°C for further assays. The 216 isolates were analysed by RFLP analysis of the 5.8S-ITS rDNA region. PCR amplification using Internal Transcribed Spacers ITS1 and ITS4 [13]. Subsequently, PCR products were digested with the following restriction enzymes: *Cfo*I, *Hae*III and *Hinf*I. The restriction fragments were checked by electrophoresis in 2% (w/v) agarose gel. Isolates sharing the same RFLP were grouped and a type was assigned for each group.

For sequence analysis of the D1/D2 domains of the 26S rDNA gene, PCR amplification was basically performed according to Kurtzman and Robnett (1998) [14] by using NL1 and NL4 primers. The PCR product was purified using UltraClean PCR Clean Up kit (MoBio, Carlsbad, CA) according to the manufacturer's instructions. Direct sequencing of the purified PCR products was performed by ABI Prism BigDye Terminator Cycle Sequence Ready Reaction Kit (Applied Biosystems, Stafford, TX). The sequences were aligned, by using the BLAST search program, with complete or nearly complete 26S rDNA gene sequences retrieved from the EMBL nucleotide sequence data libraries (http://www.ncbi.nlm.nih.gov/BLAST/[15]). The sequences determined in the course of this work were submitted to GenBank under accession numbers ranging from KF298095 to KF308283. Except where indicated, enzymes and molecular biology kits were purchased from Boehringer Mannheim GmbH, Mannheim, Germany.

2.3 Nutritional competition assay among filamentous fungi and oil-mill isolated yeasts on agar plates

A. flavus (CECT 2686) and *A. parasiticus* (CECT 2681) were obtained from the Spanish Type Culture Collection (Paterna, Spain) and maintained on malt agar plates. Spore suspensions were prepared by collecting spores from 6-day-old colonies (grown on malt agar at 28°C) in water with 0.2-mL Tween 80 added to facilitate the dispersal of conidia. The assay was basically carried out as described by Bleve et al. [11]. Briefly, a suspension of spores was spread on malt agar and yeasts were streaked onto the surface. The effect was considered positive when an inhibition zone surrounded the streak of the challenger yeast. Therefore, 12 representative strains were submitted to competition experiments on malt agar plates. Yeast cultures were maintained on slants of YPD agar. Suspensions were prepared by inoculating 5 mL of YPD with a loop of cells and incubation at 28°C for 24-48 h. The final concentration of yeasts was adjusted to spread 100 μl of 10^7 CFU. Three 10-μl of fungal spore suspensions (10^3 and 10^5 spores/ml) were separately spotted on each serial of plates. Three replicate experiments for each fungus, spotted to make it possible to measure the radial extension rates of the colonies, were performed. Plates not inoculated with yeasts were used as control. Analysis software ImageJ (http://rsb.info.nih.gov/ij/) was used to determine the growth area. Fungal growth inhibition was determined as the percent of growth area decrease compared to the control as described by Virgili et al. [16].

3. Results and discussion

3.1 Culture conditions and yeast diversity

In this work we have isolated a total of 216 yeast strains from olive fruits, olive paste and olive pomace, from three different oil mills, in two successive campaigns. Surprisingly, the variety observed in the Ontinyent and Venta del Moro oil mills was quite lower (17 and 17 isolates, respectively) when it was compared with the Villamalea mill (182 isolates). Romo-Sanchez et al. [5] reported higher variability, although with a very limited number of isolates when studied oil mills from other Spanish region. This ecological distribution can be attributed to environmental differences or even to different production processes carried out. Microbiological studies revealed that *Saccharomyces* and *Candida* species are frequently present in this ecological niche [6]. No differences were detected, regarding this aspect, between the two campaigns included in this work. After a visual inspection of colonies, we observed that 21 strains had a yellowish colony appearance, 5 reddish and 67 of them were cream. 168 of them had butyrous colony texture and 48 had mucoid colony texture.

These are the results of a screening program carried out to improve the microbial diversity, including the use of different growth media for yeast sampling (first harvest campaign, 98 isolates). As a consequence, malt agar was discarded because supported the growth of many filamentous fungi. Tributyrin and olive oil agar media were the most satisfactory, supporting the growth of around 1.6×10^7 yeasts/ml and 4.9×10^6 yeasts/ml, respectively. The total growth in other culture media was lesser. Moreover, microbial biodiversity was similar among the four media (26 to 42 different isolates were obtained). As an attempt to enhance the diversity of sample an enrichment media was also used, but no increase in this aspect was detected. As a consequence, for the second-year sampling, olive oil agar was the only culture media used for this purpose, without an enrichment step.

3.2 Identification using rDNA gene sequence

Genomic DNA was extracted, and PCR was used to generate amplicons rounding 575 bp to 850 bp of the ITS1-5.8S rDNA, which is basically on agree with Romo-Sánchez et al. [5]. RFLP of these amplicons produced a characteristic band profile for each strain that was used to make comparison with data recorded in the Yeast-ID database at the CECT web server. Then, a total of 43 representative yeast strains were subjected to PCR amplification of the D1/D2 region and were identified by comparing sequences using the NCBI blast program. According to the criteria for discrimination of yeast species proposed by Kurtzman and Robnett [17] using D1/D2 rDNA sequencing, a match greater than 99% is required to assess that an isolate of yeast is a member of a species. Following this, a total of four different genera including nine species were identified (Table 1).

Table 1. Identification and molecular characterization of 216 isolates from different substrates and olive mills

Species	Accession number (isolate type)	N° isolates	Area[a]	PCR product (bp)[b]	Restriction fragments (bp)[b]		
					*Hinf*I	*Hha*I	*Hae*III
Candida molendinolei	KF298101	6	O	650	325+325	600	600
Saccharomyces paradoxus	KF298095	7	O	800	400+400	375+350	300+250
Candida sp.	KF298105	4	O	650	325+325	650	650
Saccharomyces paradoxus	KF298099	34	VL	800	400+400	375+350	300+250
Candida molendinolei	KF298103	35	VL	650	325+325	600	600
Candida sp.	KF298106	2	VL	575	325+200	575	575
Candida adriatica	KF298107	36	VL	575	300+250	200+200+125	400+100
Candida norvegica	KF308273	41	VL	575	300+275	575	375+200
Candida boidinii	KF298109	34	VL	725	500+200	375+350	725
Candida oleophila	KF308281	12	VM	575	300+250	300+275	425+200+150
Citeromyces matritensis	KF308278	4	VM	700	450+400	700	700
Cryptococcus sp.	KF308280	1	VM	575	270+260	300+280+150+100	400+100

[a] Isolation area: O, Ontinyent; VL, Villamalea; VM, Venta del Moro
[b] Size estimated in 2% agarose gel

Yeasts commonly isolated from the Ontinyent mill were *Candida molendinolei* and *Saccharomyces paradoxus*, from Villamalea (*C. molendinolei*, *C. boidinii* and *C. norvegica*) and from Venta del Moro (*C. oleophila* and *Citeromyces matritensis*). *Cryptococcus laurentii* is the predominant species isolated from fresh olives [3]. It has been used as a yeast for biocontrol [18]. *S. cerevisiae* has been previously described as a striking strain, as it may be spoilage yeast during olive oil storage [1, 9]. However, due to the low lipolytic of the isolated strains, it is not considered as a problem. The isolation of Candida strains is interesting provided they have been proved to be able to produce high-value compounds while degrading oil mill waters (OMW) [19]. The demand for yeast strains to reduce polyphenols in OMW, before wastewater release, metabolically adapted to different oil mills, justifies the effort to carry out these studies. *Citeromyces matritensis* has been previously isolated from wine-related communities [20] but not in oil mills. *C. molendinolei* and *C. adriatica* are yeasts associated with olive oil and its by-products as has been recently reported [21]. *C. boidinii* has been frequently isolated and reported in oil mills [5, 9]. This yeast has been investigated to be used as agent for bio-treatment of OMW in Greece and Italy. *C. oleophila* has been traditionally used for mould control in vivo and in vitro assays [22]. Some strains have been used (even they have been available in the market) for postharvest biocontrol in some fruits [23]. *C. norvegica* has also been isolated, although it may be a contaminating species from soil origin. Provided we have carried out our screenings in different conditions, mills and years, it may be concluded that microbial biodiversity in eastern Spain olive oil mills is lower than in southern (warmer) areas. Moreover, manufacturing procedures are not identical in different geographic areas.

3.3 Nutritional competition assay

The twelve representative yeast isolates, previously selected in the enzymatic characterization were used in a nutritional competition assay. This test was proposed in order to select yeast strains able to affect the co-inoculated fungi when colonizing a common ecological niche. A different inhibitory activity against *Aspergillus flavus* and *A. parasiticus* was found among yeast isolates (Table 2).

Table 2. Percentage of growth reduction of *Aspergillus flavus* and *A. parasiticus* caused by yeasts isolated from olive oil environment y in a nutritional competition assay

| Species | Accession N° (isolate type) | % inhibitory activity[a] | | | |
| | | *A. flavus*[b] | | *A. parasiticus*[b] | |
		10^3	10^5	10^3	10^5
Candida molendinolei	KF298101	3.1 a	2.6 a	23.3 b	19.8 a
Saccharomyces paradoxus	KF298095	80.5 e	75.6 d	88.0 e	87.2 e
Candida sp.	KF298105	23.3 b	19.8 a	53.9 c	51.2 c
Saccharomyces paradoxus	KF298099	88.4 e	83.1 e	72.4 d	69.9 d
Candida molendinolei	KF298103	2.2 a	1.2 a	55.1 c	54.2 c
Candida sp.	KF298106	0.3 a	0.1 a	3.2 a	1.6 a
Candida adriatica	KF298107	16.9 a	14.6 a	27.4 b	25.8 b
Candida norvegica	KF308273	30.9 b	28.4 b	30.6 b	29.9 b
Candida boidinii	KF298109	73.9 d	71.2 d	85.9 e	84.1 e
Candida oleophila	KF308281	75.5 d	72.3 d	74.8 d	73.1 d
Citeromyces matritensis	KF308278	23.6 b	21.2 b	23.3 b	21.4 b
Cryptococcus sp.	KF308280	19.5 a	17.4 a	16.6 a	15.7 a

[a] different letters in the same column mean significant differences (Fisher test, $P < 0.05$)
[b] Units expressed as spores/mL

Saccharomyces paradoxus, *C. boidinii* and *C. oleophila* showed the highest inhibition and *C. molendinolei* and *Candida* sp. the lowest one. Assayed yeast inhibitory activity was affected by the concentration of fungal inoculum: the growth in the lowest fungal inoculum concentration (10^3 spores/mL) was the most inhibited by antagonistic yeasts; similar growth was observed in growth when 10^5 spores/mL were used. This supports the finding that the concentration of the fungus is a key factor for yeast effectiveness ($p<0.05$). As a consequence, in future applications of antagonistic yeasts for oil biocontrol, it will be important to keep the population of contaminating fungi as low as possible, by controlling processing procedures for high quality olive oil.

Acknowledgements This work is part of the project "Isolation of enzymes with industrial interest from food waste materials" supported by grant INV-AE112-66049 from the Universitat de València, Spain.

References

[1] Giannoutsou EP, Meintanis C, Karagouni AD (2004) Identification of yeast strains isolated from a two-phase decanter system olive oil waste and investigation of their ability for its fermentation. Bioresour Technol 93: 301–306

[2] Jones CE, Murphy PJ, Russell NJ (2000) Diversity and osmoregulatory responses of bacteria isolated from two-phase olive oil extraction waste products. World J Microbiol Biotechnol 16: 555–561

[3] Hernández A, Martín A, Aranda E, Pérez-Nevado F, Córdoba MG (2007) Identification and characterization of yeast isolated from the elaboration of seasoned green table olives. Food Microbiol 24: 346–351

[4] Panagou EZ, Schillinger U, Franz CMAP, Nychas GJE (2008) Microbiological and biochemical profile of cv. Conservolea naturally black olives during controlled fermentation with selected strains of lactic acid bacteria. Food Microbiol 25: 348-358

[5] Romo-Sánchez S, Alves-Baffi M, Arévalo-Villena M, Úbeda-Iranzo J, Briones-Pérez A (2010) Yeast biodiversity from oleic ecosystems: Study of their biotechnological properties. Food Microbiol 27: 487-492

[6] Arroyo-López FN, Querol A, Bautista-Gallego J, Garrido-Fernández A (2008) Role of yeasts in table olive production Int J Food Microbiol 128: 189–196

[7] Fernandez-González M, Espinosa JC, Ubeda J, Briones AI (2001) Yeasts present during wine fermentation: comparative analysis of conventional plating and PCR–TTGE. Syst Appl Microbiol 24: 639–644

[8] Roussos S, Zaoula N, Salih G, Tantaoui-Elaraki A, Lamrani K, Cheheb M, Hassouni H, Verhé F, Perraud-Gaime I, Augur C, Ismaili-Alaoui M (2006) Characterization of filamentous fungi isolated from moroccan olive and olive cake: Toxinogenic potential of aspergillus strains. Mol Nutr Food Res 50: 500-506

[9] Leontopoulos D, Siafaka A, Markaki P (2003) Black olives as substrate for *Aspergillus parasiticus* growth and aflatoxin B$_1$ production. Food Microbiol 20:119-126

[10] Papachristou A, Markaki P (2004) Determination of ochratoxin A in virgin olive oils of Greek origin by immunoaffinity column clean-up and high-performance liquid chromatography. Food Additives and Contaminants 21: 85-92

[11] Bleve F, Grieco G, Cozzi A, Logrieco A (2006) Visconti Isolation of epiphytic yeasts with potential for biocontrol of *Aspergillus carbonarius* and *A. niger* on grape. Int J Food Microbiol 108: 204-209

[12] Katsifas EA, Giannoutsou EP, Karagouni AD (1999) Diversity of streptomycetes among specific Greek terrestrial ecosystems Lett Appl Microbiol 29: 48–51

[13] White TJ, Bruns T, Lee S, Taylor J (1990) Amplification and direct sequencing of fungal ribosomal RNA genes for phylogenetics. In: Innis MA, Gelfand DH, Sninsky JJ, White TJ (eds) PCR Protocols. A Guide to Methods and Applications, Academic Press, San Diego, CA, USA pp. 315–322

[14] Kurtzman, C.P., Robnett, C.J. (1998). Identification and phylogeny of ascomycetous yeast from analysis of nuclear large subunit (26S) ribosomal DNA partial sequences. Antonie van Leeuwenhoek 73: 331-371

[15] Altschul SF, Madden TL, Schäffer AA, Zhang J, Zhang Z, Miller W, Lipman DJ (1997) Gapped BLAST and PSI-BLAST: a new generation of protein database search programs. Nucleic Acids Res 25:3389-3402

[16] Virgili R, Simoncini N, Toscani T, Camardo-Leggieri M, Formenti S, Battilani P (2012) Biocontrol of *Penicillium nordicum* growth and ochratoxin A production by native yeasts of dry cured ham. Toxins 4: 68-82

[17] Kurtzman CP, Robnett CJ (1998) Identification and phylogeny of ascomycetous yeast from analysis of nuclear large subunit (26S) ribosomal DNA partial sequences. Antonie van Leeuwenhoek 73:331-371

[18] Meng XH, Qin GZ, Tian SP (2010) Influences of preharvest spraying *Cryptococcus laurentii* combined with postharvest chitosan coating on postharvest diseases and quality of table grapes in storage. LWT - Food Sci Technol 43: 596-601

[19] Brozzoli V, Crognale S, Sampedro I, Federici F, D'Annibale A, Petruccioli M (2009) Assessment of olive-mill wastewater as a growth medium for lipase production by *Candida cylindracea* in bench-top reactor. Bioresour Technol 10: 3395-3402

[20] Brežná B, Zenišová K, Chovanová K, Chebeňová V, Kraková L, Kuchta T, Pangallo D (2010) Evaluation of fungal and yeast diversity in Slovakian wine-related microbial communities. Antonie van Leeuwenhoek 98: 519-29

[21] Cadez N, Raspor P, Turchetti B, Cardinali G, Ciafardini G, Veneziani G, Péter G (2012) *Candida adriatica* sp. nov. and *Candida molendinolei* sp. nov., two yeast species isolated from olive oil and its by-products. Int J Syst Evol Microbiol 62: 2296-2302

[22] El-Ghaouth A, Wilson CL, Wisniewski M (1998) Ultrastructural and cytochemical aspects of the biological control of *Botrytis cinerea* by *Candida saitoana* in apple fruit. Phytopathology 88: 282-291

[23] Molinu MG, Pani G, Venditti T, Dore A, Ladu G, D'Hallewin G (2011) Sequential application of NaHCO$_3$, CaCl$_2$ and *Candida oleophila* (isolate 13L) affects significantly *Penicillum expansum* growth and the infection degree in apples. Commun Agric Appl Biol Sci, 76: 743-750

Antifungal activity of residues from aromatic waters distilled from thyme and sage

M. Zaccardelli[*,1], **G. Roscigno**[2], **C. Pane**[1] and **E. De Falco**[2]

[1] Consiglio per la Ricerca e la Sperimentazione in Agricoltura - Centro di Ricerca per l'Orticoltura, via dei Cavalleggeri 25, 84098 Pontecagnano (SA), Italy
[2] Dipartimento di Farmacia, Università degli Studi di Salerno, Via Giovanni Paolo II 132, 84084 Fisciano (SA), Italy
*Corresponding author: e-mail: massimo.zaccardelli@entecra.it, Phone: +39 089386200

Residual waters result from the aqueous phase obtained during the extraction of essential oils from medicinal plants. They could content water-soluble volatile components conferring antimicrobial activity. In this study two residual aromatic waters, resulting from distillation of thyme and sage, were assayed for their antifungal activity against plant pathogens *Rhizoctonia solani* and *Sclerotinia minor*. To evaluate the ability of these co-products to inhibit growth of fungal plant pathogens, *in vitro* plate tests were performed. Thyme and sage residual distillation waters were separately tested by submerging in each a plug (5 mm in diameter) removed from the edge of the growing mycelia. After overnight incubation at 25 °C, the plug was transferred in the centre of a PDA Petri plates (diameter 90 mm). Control plates were prepared without the addition of the tested waters. For each treatment, plates were inoculated in triplicate and incubated at 25 °C. The diameter of the mycelia was measured daily until fungi reached the edge of control plates; data were expressed as growth rate (mm d^{-1}). Assays evidenced antimicrobial activity of residual waters that reduced the *in vitro* growth rate of the pathogens. Sage water showed highest inhibition than those of thyme. These results are promising for sustainable application in the field of plant disease control.

Keywords Plant Pathogens; Plant-derived antimicrobials

1. Introduction

The residual waters obtained during the current vapour distillation of medicinal plants (RAWs) were separated from essential oil and aromatic water. Abundant literature is available about the antimicrobial effects of essential oil and aromatic water [1,2]. On the opposite, there is a lack of studies about the potentiality of the residual waters that could contain phytochemical compounds responsible for their biological properties [3]. Moreover actually it's of great interest the recovery of by-products such as phenol from industrial processing. The bioactivity of RAWs could be explored in order to detect potential antifungal effects against plant fungal pathogens. These last are responsible of serious economic losses on different crops and, often, it is impossible to control them without synthetic fungicides. Nevertheless, the necessity to increase food security and environmental sustainability reasons, are stimulating the search of new eco-friendly and non-chemical alternatives in plant disease management [4]. The aim of this study was to investigate the antifungal activity of aromatic waters, resulting from thyme and sage distillation, for the control of the phytopatogenic fungi *Rhizoctonia solani* and *Sclerotinia minor*.

2. Materials and Methods

2.1 Fungal pathogens

Two strains of the fungal phytopathogens *Rhizoctonia solani* and *Sclerotinia minor*, were originally isolated from their specific hosts and were maintained at CRA-Centro di Ricerca per l'Orticoltura (Pontecagnano, Italy) on potato dextrose agar medium (PDA, Oxoid) at 20 °C.

2.2 Production of waters residues

Cropping residues and processing discards of sage and thyme plants (Fig. 1) were subjected to distillation in vapour-current for 3 h, according to the method recommended by Bettiol and Vincieri [5]. After the hydrodistillation, the obtained essential oils, the aromatic waters and the residues waters (RAWs) were separated one from each other. RAWs were stored at 4 °C until the use.

2.3   *In vitro* antifungal plate assay

To evaluate the antifungal activity of thyme and sage residual waters, an *in vitro* plate test was performed as described previously by Pane et al. [6]. The two plant-derived liquid formulates were separately tested by submerging them in a plug (5 mm in diameter) removed from the edge of the growing mycelia. Two concentrations, 1× and 2× (obtained reducing volume by roto-evaporator) were assayed each. After overnight incubation at 25 °C, the plug was transferred in the centre of a PDA Petri plates (diameter 90 mm). Control plates were prepared using sterile water in the place of the tested substances. For each treatment, plates were inoculated in triplicate and incubated at 25 °C. The diameter of the mycelia was measured daily until fungi reached the edge of control plates; data were expressed as growth rate (mm d^{-1}).

3. Results and Discussion

Literature survey indicates that this study provided new data about *in vitro* inhibition activity of the plant-derived RAWs against two soil-borne phytopathogens *R. solani* and *S. minor*. Assays evidenced the reduction of *in vitro* growth rate recovery of the pathogens following the antifungal treatment, with a dose-dependent effect (Fig. 2). Moreover, sage water showed highest inhibition than those of thyme. These findings indicate the potential of the assayed co-product of stem-distillation as antifungals against phytopathogens. It's an interesting result, since demand for natural products is increasing in agrochemical industry in order to find effective alternatives to synthetic fungicides. The exploration of RAWs opens perspectives in this direction. However, further investigation is still necessary to search determinants of pathogen suppressiveness through the chemical characterization of the most effective RAWs and their proof on *in-vivo* systems.

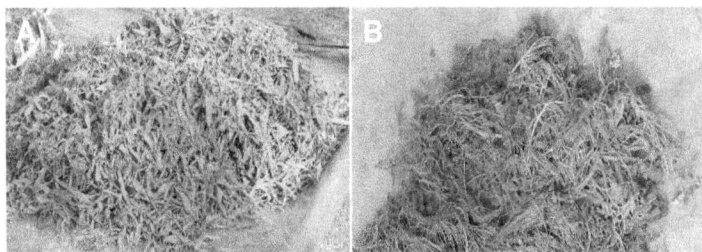

Fig. 1   Processing solid residues of sage before (A) and after (B) stem-distillation.

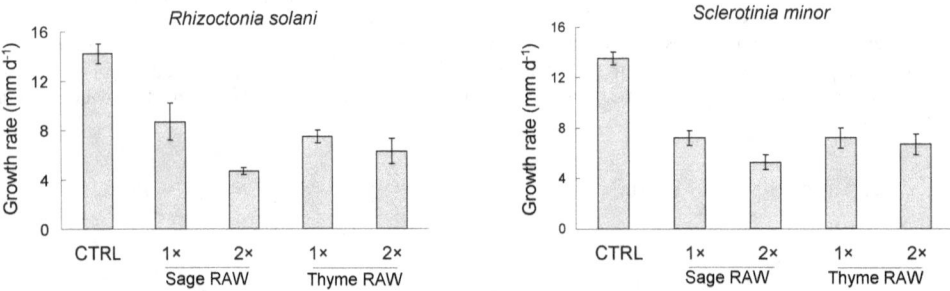

Fig. 2   Effect of mycelial plug treatments with sage and thyme RAWs at 1× or double-concentrated (2×) on the rate of growth recovery of the phytopathogenic fungi *R. solani* and *S. minor* compared to the untreated control (CTRL).

References

[1]   Ciccarelli D, Noccioli C, Pistelli L. Chemical composition of essential oils and aromatic waters from different Italian *Anthemis maritima* Populations. Chemistry & Biodiversity, 2013; 10:1667-1682.
[2]   Cerchiara T, Blaiotta G, Straface VS, Belsito E, Liguori A, Luppi B, Bigucci F, Chidichimo G. Biological activity of *Spartium junceum* L. (*Fabaceae*) aromatic water. Natural Resources, 2013; 4:229-234.
[3]   Almela L., Sánchez-Muñoz B., Fernández-López J.A., Roca M.J., Rabe V., 2006. Liquid chromatograpic-mass spectrometric analysis of phenolic and free radical scavenging activity of rosemartry extract from different raw material. Journal of Chromatograòhy, 1120, 221:229.

[4]   van Lenteren JC. A greenhouse without pesticides: fact or fantasy? Crop Protection, 2000; 19: 375-384

[5]   Bettiol F, Vincieri F. Manuale delle Preparazioni Erboristiche. Fitoterapici, fitocosmetici, prodotti erboristici, integratori alimentari a base di piante. Tecniche Nuove (Ed.), 2009.

[6]   Pane C, Villecco D, Roscigno G, De Falco E, Zaccardelli M. Screening of plant-derived antifungal substances useful for the control of seedborne pathogens. Archives of Phytopathology and Plant Protection, 2013; 46:1533-1539.

Antifungal activity of wild *Capsicum* foliar extracts containing polyphenols against the phytopathogens *Alternaria alternata, Rhizoctonia solani, Sclerotinia minor* and *Verticillium dahliae*

C. Pane[*,1], **F. Fratianni**[2], **M. Caputo**[1], **M. Parisi**[1], **F. Nazzaro**[2] and **M. Zaccardelli**[1]

[1] Consiglio per la Ricerca e la Sperimentazione in Agricoltura - Centro di Ricerca per l'Orticoltura, via dei Cavalleggeri 25, 84098 Pontecagnano (SA), Italy
[2] Consiglio Nazionale delle Ricerche, Istituto di Scienze dell'Alimentazione, via Roma 52, I-83100 Avellino, Italy
*Corresponding author: e-mail: catello.pane@entecra.it, Phone: +39-089-386211

Natural plant-derived antifungals can be explored as valid alternative to conventional chemical fungicides for preventing plant pathogens development. In this study, thirteen aqueous crude foliar extracts of *Capsicum* genotypes were examined for their antifungal activity. Grinded 70 °C-dried leaves were autoclaved, at a dose of 0.5 g ml^{-1}, in potato dextrose agar medium and assayed for growth of phytopathogenic fungi on Petri dishes. The employed fungal pathogens were *Alternaria alternata, Rhizoctonia solani, Sclerotinia minor* and *Verticillium dahlae*. Growth rate of all tested fungi were significantly suppressed by extracts obtained from wild *Capsicum annuum, C. annuum* var. *glabriumsculum, C. baccatum* var. *baccatum, C. chinense* and *C. tovarii*. Microscopic analysis revealed remarkable morphological alterations in hyphal structures of samples exposed to the most bioactive leaf extract. UPLC–MS/MS analysis was performed to obtain polyphenolic profiles of extracts and quantify the individual known components, including gallic acid, chlorogenic acid, catechin, caffeic acid, epicatechin, coumaric acid, rutin, ferulic acid, hyperoside, luteolin, and quercetin. The role of polyphenols in extract antimicrobial activity has been discussed.

Keywords Plant Pathogens; Plant-derived antimicrobials

1. Introduction

Toxicological concerns associated to the use of chemical fungicides in agriculture and the increased awareness about food security are stimulating the general interest on the use of natural substances in controlling plant pathogens. The exploitation of phytochemical compounds extracted from plants, may be a promising and valid alternative to conventional disease management [1]. Plants produce a wide range of biologically active compounds that are, in majority, secondary metabolites that serve as plant defense mechanisms against biotic stresses. They can be extracted, either as crude or purified, and used as potential pathogens control agents [2]. Various plant crude extracts with antimicrobial efficacy, in fact, were found toxic to several phytopathological fungi [3]. Antifungal traits of plant extracts are due to their content in different classes of bioactive chemicals such as flavonoids, phenols, tannins, alkaloids, quinons, saponins and sterols [4].

Literature surveys indicated that medicinal plants are largely investigated for antifungal properties. Few, instead, is developed from cultivated plants and associated wild genotypes. Some rare case comes, for example, from garlic extracts that have been successfully tested against various fungal pathogens [5]. The genus *Capsicum* L. is a member of *Solanaceae* family worldwide extensively cultivated, that originated from Central and South America and include a complex of five domesticated species, such as *C. annuum* L., *C. frutescens* L., *C. baccatum* L. var. *pendulum* and *C. pubescens* R. & P. [6]. Cultivated and wild genotypes are characterized for great metabolite biodiversity [7]. Nevertheless, little attention has been given to involvement of their antimicrobial properties in the control of plant pathogens.

The present work is aimed to explore: (i) the antifungal activity of foliar extracts of a set of *Capsicum* species against *Alternaria alternata, Rhizoctonia solani, Sclerotinia minor* and *Verticillium dahliae*; (ii) total polyphenol profiles of the extracts and (iii) hypothetical mechanisms governing antifungal activity.

2. Materials and Methods

2.1 Fungal pathogens and plant materials

The fungal pathogens used in this study were *Alternaria alternata, Rhizoctonia solani, Sclerotinia minor* and *Verticillium dahliae*, maintained at CRA-Centro di Ricerca per l'Orticoltura (Pontecagnano, Italy). Fungi were maintained on potato dextrose agar medium (PDA, Oxoid) at 20 °C.

Capsicum accessions used in this study are listed in Table 1. They were growth in pots under greenhouse system and, at fruit ripening, leaves were collected, dried at 70 °C and grounded until to obtain a powder.

Table 1 List of *Capsicum* accessions collected from germplasm banks of the Centre for Genetic Resources of the Netherlands (CGN) and IPK Gatersleben (IPK) used in this study.

Sample	Germplasm bank origin	Germplasm bank N°	Pedigree[a]	Botanical name
1	CGN	CGN 21526	W	*C. annuum*
2	IPK	CAP1225	NA	*C. annuum*
3	IPK	CAP539	NA	*C. annuum* var. *glabriumsculum*
4	CGN	CGN 21467	W	*C. baccatum* var. *baccatum*
5	CGN	CGN21512	L	*C. baccatum* var. *pendulum*
6	CGN	CGN22834	L	*C. baccatum* var. *pendulum*
7	CGN	CGN23208	W	*C. chacoense*
8	CGN	CGN17220	NA	*C. chinense*
9	IPK	CAP1546	NA	*C. eximium*
10	CGN	CGN22792	W	*C. frutescens*
11	CGN	CGN22794	W	*C. praetermissum*
12	CGN	CGN22796	L	*C. pubescens*
13	CGN	CGN22876	W	*C. tovarii*

[a]Legend: W=wild; L=Landrace; NA=not available

2.2 *In-vitro* plate assay and light microscopic observations

The antifungal activity of plant materials was investigated through the *in-vitro* assay described by Zaccardelli et al. [8] with slight modification. *Capsicum* meals, at a rate of 0.5 g ml^{-1}, were submitted to an autoclaving water extraction process directly in amended PDA medium (0.1×). Then, Petri dishes (diameter 90 mm) were poured, inoculated in the centre with a plug (5 mm in diameter) taken from growing fungal colony and incubated at 25 °C. The diameter of the mycelia was measured daily until fungi reached the edge of not-amended plates used as control. Data were expressed as growth rate (mm d^{-1}).

Data from the plate inhibition assays of the fungi with extracts were processed with the analysis of variance (ANOVA) and means were separated with Duncan's test.

One mycelium plug of each fungus was placed on PDA (0.1×) plate, at 3-cm distance from a well (0.5 cm) filled with 100 μL of wild *Capsicum annuum* (CGN21526) antifungal extract prepared as indicate above. After two days, samples of mycelium were taken from the edge of the fungal colony expanding in the phytochemical-diffusion area, stained with trypan blue and examined under light microscope to record structural abnormalities. The observations were compared with a not-treated control.

2.3 Analysis of polyphenols

The total phenolic contents were determined following the method of Singleton and Rossi [9] using the Folin–Ciocalteu phenol reagent. The absorbance at k = 760 nm was determined at room temperature using a Cary UV/Vis spectrophotometer (Varian, Palo Alto, CA, USA). Quantification was based on a standard curve generated using gallic acid. The results were expressed as μg of gallic acid equivalents (GAE)/ml of fresh weight (fw) of the product ± standard deviation (SD).

An ACQUITY Ultra Performance LC™ system (Waters, Milford, MA, USA) linked to a PDA 2996 photodiode array detector (Waters) was used for ultra high-performance liquid chromatography (UPLC) analyses. Empower software was used to control the instruments and for data acquisition and processing. The extracts and the standards (previously dissolved in methanol) were filtered (0.45 μm, Waters, Milford, MA, USA) before analysis. The analyses were performed at 30 °C using a reversed phase column (BEH C18, 1.7 μm, 2.1 • 100 mm, Waters) following the method of Fratianni et al. [10]. The mobile phase consisted of solvent A (7.5 mM acetic acid) and solvent B (acetonitrile) at a flow rate of 250 μL min^{-1}. Gradient elution was employed, starting with 5% B for 0.8 min, then 5–20% B over 5.2 min, isocratic 20% B for 0.5 min, 20–30% B for 1 min, isocratic 30% B for 0.2 min, 30–50% B over 2.3 min, 50–100% B over 1 min, isocratic 100% B for 1min, and finally 100–5% B over 0.5 min. At the end of this sequence, the column was equilibrated under the initial conditions for 2.5 min. The pressure ranged from 6000 to 8000 psi during the chromatographic run. The effluent was introduced into an LC detector (scanning range: 210–400 nm, resolution: 1.2 nm). The injection volume was 5 μL.

2.4   Free radical scavenging capacity

The free radical scavenging activity of the extracts was determined using 2,2-diphenyl-1-picrylhydrazyl (DPPH) [11]. The analysis was performed in microplates by adding 7.5 μL of extract to 303 μL of a methanolic DPPH solution (6×10^{-5} M). Subsequently, the absorbance at k = 517 nm was read (Cary 50 MPR, Varian Italia, Cernusco sul Naviglio, Milano, Italy). The absorbance of DPPH without antioxidant (control sample) was used for baseline measurements. The scavenging activity was expressed as the 50% Inhibitive concentration (IC50), which was defined as the sample concentration (mg) necessary to inhibit the DPPH radical activity by 50% during a 60-min incubation period. These experiments were performed in triplicate, and the results were expressed as mean value ± standard deviation.

3. Results and Discussion

3.1   Antifungal activity

Capsicum meals inhibited growth of very feared plant pathogenic fungi at variable levels. The highest suppressiveness was showed by meal from *Capsicum annuum* wild accession CGN21526. In addition, extracts of *C. annuum* var. *glabriumsculum*, *C. baccatum* var. *baccatum*, *C. chinense* and *C. tovarii*, also proved significantly suppressive. *R. solani* and *V. dahliae* resulted significantly sensitive to all extracts at dose employed. *S. minor* was also significantly inhibited by crude extracts of *C. eximium*, *C. chacoense* and *C. praetermissum*. While, *A. alternata* was also suppressed by *C. eximium*, *C. pubescens* and *C. frutescens* extracts.

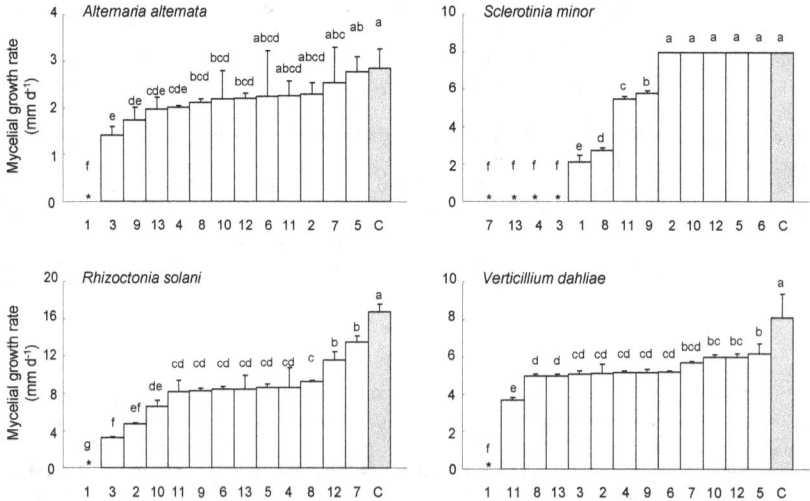

Fig. 1   Effect of the *Capsicum* extracts listed in Table 1, on radial growth rate of *A. alternata*, *R. solani*, *S. minor* and *V. dahliae*, compared to the control (C). Lowercase letters indicate differences statistically significant.

Accordingly to our findings, Soumya and Nair [12] produced aqueous extracts of *C. frutescens* leaves and tested for its antifungal activity against *Aspergillus flavus*, *A. niger*, *Penicillium* sp., *Rhizopus* sp.; Kunasakdakul and Suwitchayanon [13], instead, have documented that chili pepper extract had antifungal activity against *Alternaria brassicola*. The most bioactive pepper extract (wild *C. annuum* CGN21526) affected the viability of hyphae on treated fungi, as showed by light microscopy. Microscopic analysis revealed remarkable morphological alterations in hyphal structures of samples exposed to the bioactive extract. These fungi displayed, in general, morphological alterations with shrunken and collapsed forms characterized by severe shrinking of hyphae, subsequently to the loss of cell turgidity, internal discolorations and cytoplasmic coagulation and the formation of irregular shapes. Particular effects were showed by *R. solani* that additionally showed swollen bodies on some branch, while a serpentine shape of hyphae was displayed by *V. dahliae*.

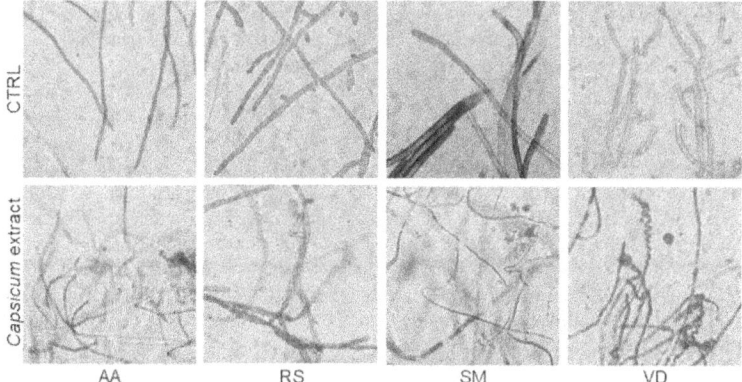

Fig. 2 Light microscopic observations of fungal hiphae of *A. alternata* (AA), *R. solani* (RS), *S. minor* (SM) and *V. dahliae* (VD) exposed to wild *C. annuum* (CGN21526) extract, compared to control (CTRL).

In a previous study, shrivelling, blistering and autolysis of hyphal structures were also documented on *Sclerotinia sclerotiorum* treated with oregano and fennel essential oils [14], while, *R. solani* were completely collapsed under treatment with *Desmos chinensis* extracts, likely due to alteration in the membrane permeability [15]. Actually, the interference with the phospholipids bilayer of the cell membrane was reported as one of the mechanisms of action described for phytochemicals [16].

Table 2 Antioxidant activity assessed by DPPH˙ radical, total polyphenol content expressed as gallic acid equivalents (GAE) and amount of each known polyphenol in *Capsicum* foliar extracts listed in Table 1.

Extracts	IC50 DPPH (μl ml^{-1})	Total polyphenols (μg GAE ml^{-1})	Known polyphenols (μg ml^{-1})										
			Gallic Acid	Chlorogenic Acid	Catechin	Caffeic Acid	Epicatechin	Coumaric Acid	Rutin	Ferulic Acid	Hyperoside	Luteolin	Quercetin
1	6.639	402.745	27.641	53.538	16.744	51.972	-	-	136.306	8.822	36.062	-	-
2	6.167	535.244	21.837	32.863	8.978	20.110	-	-	84.028	28.485	-	26.500	26.978
3	4.703	463.574	14.846	59.800	36.613	13.075	78.583	7.135	153.956	24.966	25.232	-	-
4	6.591	336.218	28.807	49.892	16.626	78.579	-	-	-	10.673	74.956	-	-
5	6.627	321.53	20.473	40.125	9.979	32.854	-	-	112.244	7.852	112.128	-	-
6	3.53	595.769	24.703	162.842	-	70.713	-	-	-	34.951	67.669	-	16.612
7	14.143	329.647	19.192	188.216	13.523	51.507	-	-	-	8.801	33.405	-	-
8	9.37	318.78	22.195	159.984	8.327	12.907	41.788	-	134.809	8.963	30.531	-	-
9	8.77	343.357	18.920	204.054	10.296	41.939	54.982	8.304	143.633	5.252	28.767	-	-
10	14.321	214.56	30.180	50.534	9.918	27.716	-	-	113.592	5.429	15.136	-	-
11	7.869	338.4	27.183	108.370	5.202	46.804	57.073	-	151.657	8.191	34.335	-	-
12	8.963	300.43	20.033	90.920	-	46.141	-	-	-	14.387	28.607	-	-
13	8.007	337.263	25.059	117.277	18.513	40.728	-	-	109.231	8.208	35.275	-	-

3.2 Putative role of polyphenols

To investigate hypothetical mechanisms governing the fungal inhibition, polyphenolic compounds contained in the extracts were analysed. Total phenolic content, DPPH free radical scavenging activity and quantitative analysis of the known polyphenols, were showed in Table 2. Gallic acid, chlorogenic acid, catechin, caffeic acid, coumaric acid, rutin, ferulic acid, epicatechin, hyperoside, luteolin and quercetin, were the known phenolic compounds identified. Polyphenols are a major family of secondary metabolites that occur in a wide range of plant species well known for their antifungal activity [17]. Here, these molecules may be responsible for the bioactivity of extracts. The mechanisms underlying antimicrobial properties of phenolic compounds concern the ability to alter the functionality of biological membranes [18]. The morphological damages observed on hyphae due to the toxic extract could be compatible with losses of cell membrane integrity.

3.3 Conclusions

This study showed that the most active and fungitoxic *Capsicum* material may be an excellent candidate for future development of bio-fungicides. Moreover, the adopted extraction methodology, which does not use chemical solvents and, however, ensures the sanitation of plant material through hot sterilization, accord the concept of agricultural sustainability. Aqueous extraction method may be readily used in practice to improve eco-friendly strategy of crop disease management through a self-supply of derived bioactive substances utilizing plant residues. Therefore, findings of the present work suggest that plant-derived products with antifungal properties could have a potential for practical applications. However, future works about applicative aspects of *Capsicum* extract-based formulates in plant disease control, are necessary.

Acknowledgements   This work was supported by the Italian Ministries of Education, University and Research and of Economic, through the funding programme EU P.O.N. Research and Competitiveness 2007-2013 under Grant No. PON_02_00395_3215002, with the project "Valorizzazione di produzioni ortive campane di eccellenza con strumenti di genomica avanzata (GenHort)".

References

[1]   Gahukar RT. Evaluation of plant-derived products against pests and diseases of medicinal plants: A review. Crop Protection, 2012; 42:202-209.
[2]   Da Cruz Cabral L, Fernández Pinto V, Patriarca A. Application of plant derived compounds to control fungal spoliage and mycotoxin production in foods. International Journal of Food Microbiology. 2013; 166:1-14.
[3]   Ishnava KB, Chauhan KH, Bhatt CA. Screening of antifungal activity of various plant leaves extracts from Indian plants. Archives of Phytapology and Plant Protection, 2012; 45:152-160.
[4]   Arif T, Mandal TK, Dabur R. Natural products: Anti-fungal agents derived from plants. Opportunity, challenge and scope of natural products in medicinal chemistry, 2011pp. 283-311.
[5]   Sealy R, Evans MR, Rothrock C. The effect of a garlic extract and root substrate on soilborne fungal pathogens. HortTechnology, 2007; 17:169-173.
[6]   Manzur JP, Penella C, Rodríguez-Burruezo A. Effect of the genotype, developmental stage and medium composition on the *in vitro* culture efficiency of immature zygotic embryos from genus *Capsicum*. Scientia Horticulturae, 2013; 161:181–187.
[7]   Wahyuni Y, Ballester AR, Sudarmonowati E, Bino RJ, Bovy AG. Metabolite biodiversity in pepper (*Capsicum*) fruits of thirty-two diverse accessions: Variation in health-related compounds and implications for breeding. Phytochemistry, 2011; 72:1358–1370.
[8]   Zaccardelli M, Campanile F, Cammareri M, Grandillo S. Agronomical use of α-Tomatine and crude extracts of *Solanum* spp. to control phytopathogenic fungi. Acta Horticulturae, 2011; 914:401-404.
[9]   Singleton VL, Rossi JA. Colorimetry of total phenolics with phosphomolybdic-phosphotungstic acid reagents. American Journal of Enology and Viticulture, 1965; 16:144–158.
[10] Fratianni F, Coppola R, Nazzaro F. Phenolic composition and antimicrobial and antiquorum sensing activity of an ethanolic extract of peels from the apple cultivar Annurca. Journal of Medicinal Food, 2011; 14:957–963.
[11] Brand-Williams W, Cuvelier ME, Berset C. Use of a free radical method to evaluate antioxidant activity. LWT – Food and Science Technology, 1995; 28:25–30.
[12] Soumya SL, Nair BR. Antifungal efficacy of *Capsicum frutescens* L. extracts against some prevalent fungal strains associated with groundnut storage. Journal of Agricultural Technology, 2012; 8:739-750.
[13] Kunasakdakul K, Suwitchayanon P. Antimicrobial activities of chili and black pepper extracts on pathogens of *Chinese kale*. CMU Journal of Natural Science, 2012; 11:135-141.
[14] Soylu S, Yigitbas H, Soylu EM, Kurt S. Antifungal effects of essential oils from oregano and fennel on *Sclerotinia sclerotiorum*. Journal of Applied Microbiology, 2007; 103:1021–1030.
[15] Plodpai P, Chuenchitt S, Petcharat V, Chakthong S, Voravuthikunchai SP. Anti-*Rhizoctonia solani* activity by *Desmos chinensis* extracts and its mechanism of action. Crop Protection, 2013; 43:65-71.
[16] Sánchez E, García S, Heredia N. Extracts of edible and medicinal plants damage membranes of *Vibrio cholerae*. Applied Environmental Microbiology, 2010; 76:6888–6894.
[17] Dambolena JS, López AG, Meriles JM, Rubinstein HR, Zygadlo JA. Inhibitory effect of 10 natural phenolic compounds on *Fusarium verticillioides*. A structure–property–activity relationship study. Food Control, 2012; 28:163-170.
[18] Gurjar M, Ali S, Akhtar M, Singh K. Efficacy of plant extracts in plant disease management. Agricultural Sciences, 2012; 3:425-433.

Antileishmanial activity of low molecular weight chitin prepared from shrimp shell waste

Rym Salah-Tazdaït[1,2*], Djaber Tazdaït[1,2], Zoubir Harrat[3], Naouel Eddaikra[3], Farida Moulti-Mati[1], Nadia Abdi[2] and Nabil Mameri[4]

[1] Laboratoire de biochimie appliquée et biotechnologies (LaBAB), Mouloud MAMMERI University of Tizi-Ouzou, Algeria
[2] Unité de Recherche en Ingénierie et Environnement (URIE), National Polytechnics School, Algiers, Algeria
[3] Institut Pasteur d'Algérie (IPA), Algiers, Algeria
[4] Département Génie chimique, University of Technology of Compiègne, France
* Corresponding author: rymsalah4@gmail.com

Chitin is a tough, protective, semitransparent substance, primarily a nitrogen-containing polysaccharide, forming the principal component of the shell of crustacean, cuticles of insects and cell walls of fungi. The waste of this natural polymer is a major source of surface pollution in coastal areas. It has been proved to be biologically renewable, biodegradable, biocompatible, non-antigenic, non-toxic and biofunctional. In the present study, chitin was chemically extracted from shrimp shells. The obtained chitin was depolymerized by HCl to prepare low molecular weight chitin. Then, chitin and low molecular weight chitin were characterized. Further, antileishmanial activity of low molecular weight chitin was evaluated using *Leishmania infantum* LIPA 137 and *Leishmania infantum* LIPA 155/10, two reference strains isolated from patients in Pasteur institute from Algeria. The results showed effective antileishmanial activity of low molecular weight chitin against *Leishmania infantum* LIPA 137, but no antileishmanial activity of low molecular weight chitin against *Leishmania infantum* LIPA 155/10. It was also demonstrated that *Leishmania infantum* LIPA 155/10 is resistant to leishmaniasis drug glucantime® and *Leishmania infantum* LIPA 137 is sensitive to glucantime®. Further studies are necessary to determine the in vivo activities and applications of chitin and derivatives, in particular, in the design of new lines of drugs for use in the treatment of leishmaniasis and hopefully eradication.

Keywords antileishmanial; chitin; *Leishmania infantum*; shrimp shell waste

1. Introduction

Morbidity and mortality because of the leishmaniasis, a parasite disease, cause an estimated 2.4 million disability-adjusted life-years. Globally, there are 1.5–2 million new cases estimated and 70 000 deaths each year, and 350 million people are at risk of infection and disease [1]. In northern Africa, Algeria is one of the eight countries that constitute 90% of cutaneous leishmaniasis in the World. Leishmaniasis contributes significantly to the propagation of poverty, because treatment is expensive and hence either unaffordable or it imposes a substantial economic burden, including loss of wages. Leishmaniasis is endemic in 88 countries on five continents. Surveillance data indicate that the global number of cases has increased during the past decade and the emergence of antileishmanial drug resistance. To date, there is no vaccine in routine use against leishmaniasis [2]. The genus Leishmania is parasitic protozoa responsible for the leishmaniasis, a group of diseases affecting human and various animal populations. Co-infection leishmaniasis/HIV is a fatal synergy characterized by both infections mutually reinforcing their impact on the immune system. The major clinical syndromes found in human beings are cutaneous, mucocutaneous and visceral leishmaniasis, but these can present in a wide variety of forms [3].

On the other hand, much research has focused on chitin and derivatives as a source of bioactive material during past few decades [4, 5, 6]. Chitin is a linear polysaccharide joined by β-(1,4)-linked N-acetylglucosamine (GlcNAc) units [7]. It is the second most abundant natural polymer after cellulose [8]. Their unique properties, biodegradability, biocompatibility and non-toxicity, make them useful for a wide range of applications. Although chitin has very strong functional properties in many areas, the water-insoluble property of α-chitin is disadvantageous for its wide application [9]. In the research field of chitin, functional property has been developed for pharmaceutical and new drug candidate [5, 6, 10].

The purpose of this work is the determination of the antiparasite activities of low molecular weight chitin using *Leishmania infantum* LIPA 137 and *Leishmania infantum* LIPA 155/10, two reference strains isolated from patients in Pasteur institute from Algeria. Up to now, chitin has not been used as an antileishmanial active drug against *Leishmania infantum* strain.

2. Materials and methods

All chemicals used in this study were analytical grade and purchased from Sigma Chemical Co. (St. Louis, MO).

2.1 Test materials

Shrimp shells were obtained from a seafood restaurant. It was confirmed that all shells were from a single species of shrimp *Parapenaeus longirostris* (Lucas, 1846).

2.1.1 Chitin extraction

For extraction of chitin, the shrimp shells were washed, boiled and ground into a fine powder. After that, the powder was demineralized, deproteinized, filtrated, washed and dried [5].

2.1.2 Low molecular weight chitin preparation

Low molecular weight chitin was obtained by hydrolysis of chitin. For that, 1 g chitin was hydrolysed by 50 ml 7 N HCl at 70 °C during 3 h [5].

2.2 Analytical methods

Dried chitin and low molecular weight samples (1 mg) were dispersed, separately, in 100 mg of anhydrous KBr and pressed. The IR spectra were recorded at room temperature in the wavenumber range of 400–4000 cm^{-1} and referenced against air with a Nicolet 380 FTIR instrument (Thermoelectron Corporation).

Average molecular weight of low molecular chitin was estimated by FPLC which incorporated an Amersham Bioscience Instrument and a SuperdexTM 200 10/300 GL TricornTM high performance column.

The viscosity averaged molecular weights of chitin was determined using the Mark–Houwink equation:

$$\eta = K(M_v)^a \qquad (1)$$

where η is the intrinsic viscosity of chitin, K and a are constants that depend on the polydispersity of chitin and the solvent system used, and M_v is the viscosity averaged molecular weights.

The values of constants were previously determined to be K = 0.24 cm^3/g and a = 0.69 at 25 °C for chitin [11].

HPAEC analyses were carried out on a Dionex ICS-3000 system.

^{13}C NMR and ^{1}H NMR analyses were achieved with a spectrometer equipped with ^{13}C/^{1}H dual probe.

2.3 Antiparasite assay

Antiparasite activities of chitin and low molecular weight chitin were evaluated using two reference strains, *Leishmania infantum* LIPA 137 and *Leishmania infantum* LIPA 155/10.

Strain *Leishmania infantum* LIPA 137 was obtained from Pasteur Institute of Algeria. *Leishmania infantum* LIPA 137 is a strain sensitive to Glucanthime®. Toxicity of chitin and low molecular weight chitin against *Leishmania infantum* promastigotes was assessed as previously described by Gosland et al. (1989) [12] after some modifications.

3. Results and discussion

Synthesis of chitin and low molecular weight chitin were confirmed by FT-IR data as follows (Fig. 1).

Chitin hydrolysis was established by similarities of chitin characteristics picks. There is also deacetylation, caused by high concentrations of HCl, and confirmed by decreasing of CH$_3$ functional groups (1376.5 cm^{-1}), decreasing of CO functional groups (1187.5 cm^{-1} and 1401.0 cm^{-1}), and increasing of NH functional groups (563.0 cm^{-1}).

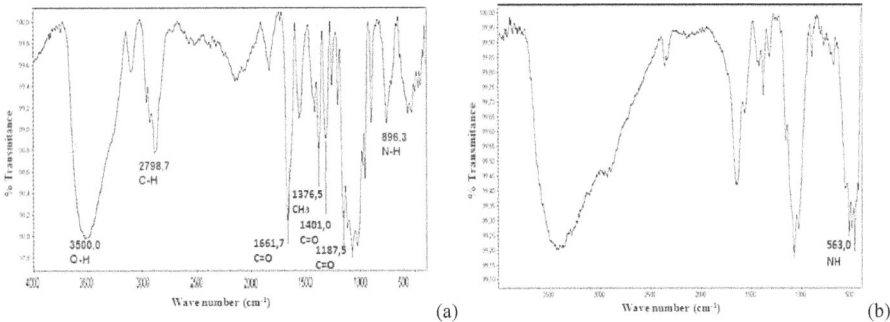

Fig. 1 FT-IR spectra of chitin (a) and low molecular weight chitin (b).

Hydrolysis of chitin permitted the production of a low molecular weight chitin and not oligosaccharides. This observation was confirmed by the absence of characteristic picks of oligosaccharides in both ^{13}C NMR plot (Fig. 2a) and HPAEC plot (Fig. 2b). Also, there is no production of glucosamine or N acetyl glucosamine as illustrated by ^{1}H NMR plot (Fig. 3).

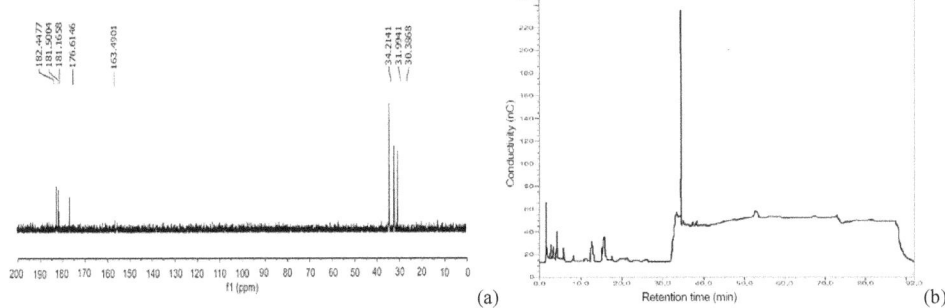

Fig. 2 ^{13}C NMR (a) and HPAEC (b) spectra of low molecular weight chitin filtrate.

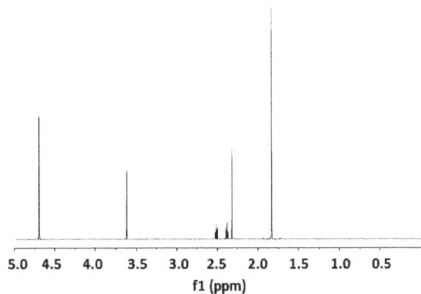

Fig. 3 ^{3}H NMR spectrum of low molecular weight chitin filtrate.

The cytotoxic effect of chitin on *Leishmania infantum* LIPA 137 (glucantime® sensitive) was evaluated. The results (Fig. 4) indicate that chitin has the potential to suppress 100% of promastigotes growth at concentrations equal or superior to 5000µg/ml. Also, the cytotoxic effect of low molecular weight chitin on *Leishmania infantum* LIPA 137 was evaluated. The results (Fig. 4) indicate that low molecular weight chitin has the potential to suppress 100% of promastigotes growth at concentrations equal or superior to 1000µg/ml.

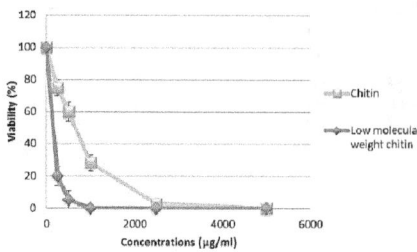

Fig. 4 Antiparasite activity of chitin and low molecular weight chitin with various concentrations against *Leishmania infantum* LIPA 137 strain.

The cytotoxic effect of chitin on *Leishmania infantum* LIPA 155/10 (glucantime® resistant) was evaluated. The results (Fig. 5) indicate that chitin exhibited no cytotoxic effects at concentrations inferior or equal to 5000µg/ml. Similarly, the cytotoxic effect of low molecular weight chitin on *Leishmania infantum* LIPA 155/10 was evaluated. The results (Fig. 5) indicate that low molecular weight chitin exhibited no cytotoxic effects at concentrations inferior or equal to 5000µg/ml.

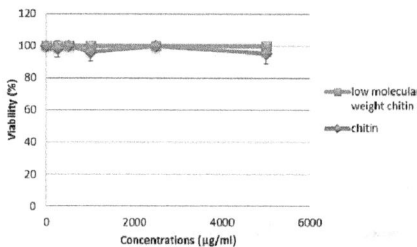

Fig. 5 Antiparasite activity of chitin and low molecular weight chitin with various concentrations against *Leishmania infantum* LIPA 155/10 strain.

The lowest IC_{50} value was 202 µg/ml for low molecular weight chitin (Table 1). These results suggest that parasite suppression increases significantly with the decrease of the molecular weight.

Table 1 Inhibitory Concentrations 50 for chitin and low molecular weight chitin.

Compounds	IC50 (µg/ml)	Molecular weights (Da)
Low molecular weight chitin	202	2480
Chitin	600	338000

Previous work has shown that, incubation of promastigotes with chitin microparticles indicated no considerable reduction in the number of viable *Leishmania major*, suggesting of the nontoxicity of chitin microparticles [13]. This study hasn't given a hypothesize on the mechanism of the inhibitory effect observed. Further studies to determine the in vivo activity and the application of chitin and derivatives, in the design of new lines of drugs in leishmaniosis treatment will be of great interest. Indeed, chitin and low molecular chitin could be decent alternatives to other drugs described in the literature [14, 15].

4. Conclusion

Low molecular weight chitin is an attractive target for selective antiparasite drug development. In fact, low molecular weight chitin prepared from the white shrimp *Parapenaeus longirotris* (Lucas, 1846) showed a great and specific antiparasite effect on *Leishmania infantum* LIPA 137 strain.

Thus, low molecular weight chitin has promising roles in natural leishmaniasis prevention and treatment. Further studies are necessary to determine the *in vivo* activities and applications of chitin and low molecular weight chitin, in particular, in the design of new lines of drugs for use in the treatment of leishmaniasis and hopefully eradication.

References

[1] WHO. The world health report 2004. Changing history. Geneva: WHO, 2004. http://www.who.int/whr/2004/en/index.html (accessed June 12, 2007).

[2] WHO. Control of the leishmaniases. Report of a meeting of the WHO Expert Committee on the Control of Leishmaniases, 22–26 March 2010, Geneva 5-88.

[3] Reithinger R, Dujardin JC, Louzir H, Pirmez C, Alexander B, Brooker S. Cutaneous leishmaniasis. The lancet infectious desease. 2007; 7:581-96.

[4] Pires CTGVMT, Vilela JAP, Airoldi C. The effect of chitin alkaline deacetylation at different condition on particle properties. Procedia Chemistry. 2014; 9:220-5.

[5] Salah R, Michaud P, Mati F, Harrat Z, Lounici H, Abdi N, Drouiche N, Mameri N. Anticancer activity of chemically prepared shrimp low molecular weight chitin evaluation with the human monocyte leukaemia cell line, THP-1. International Journal of Biological Macromolecules. 2012; 52:333-9.

[6] Benhabiles MS, Salah R, Lounici H, Drouiche N, Goosen MFA, Mameri N. Antibacterial activity of chitin, chitosan and its oligomers prepared from shrimp shell waste. Food Hydrocolloids. 2012; 29:48-56.

[7] Alves NM, Mano JF. Chitosan derivatives obtained by chemical modifications for biomedical and environmental applications. International Journal of Biological Macromolecules. 2008; 43:401-14.

[8] Khan TA, Peh KK, Chang HS. Reporting degree of deacetylation values of chitosan: the influence of analytical methods. Journal of Pharmacy and Pharmaceutical Sciences. 2002 ; 5:205-12.

[9] Je JY, Cho YS, Kim SK. Cytotoxic activities of water-soluble chitosan derivatives with different degree of deacetylation. Bioorganic & Medical Chemistry Letters. 2006; 16:2122-6.

[10] Khoushab F, Yamabhai M. Chitin Research Revisited. Marine Drugs. 2010; 8:1988-2012.

[11] Chang KLB, Lee J, Fu WR. HPLC Analysis of N-acetyl-chito-oligosaccharides during the acid hydrolysis of chitin. Journal of Food and Drug Analysis. 2000; 8:75–83.

[12] Gosland MP, Lum BL, Sikic BI. Reversal by cefoperazone of resistance to etoposide, doxorubicin, and vinblastine in multidrug resistant human sarcoma cells. Cancer Research. 1989; 49:6901–5.

[13] Dehghani F, Hoseini MHM, Memarnejadian A, Yeganeh F, Rezaie AM, Khaze V, Sattari M, Tamijani HD, Labibi F, Mossaffa N. Immunomodulatory activities of chitin microparticles on Leishmania major-infected murine macrophages. Archives of Medical Research. 2011; 42:572–6.

[14] Abbassi F, Raja Z, Oury B, Gazanion E, Piesse C, Sereno D, Nicolas P, Foulon T, Ladram A. Antibacterial and leishmanicidal activities of temporin-SHd, a 17-residue long membrane-damaging peptide. Biochimie. 2013; 95:388–99.

[15] Athanasiou LV, Saridomichelakis MN, Kontos VI, Spanakos G, Rallis TS. Treatment of canine leishmaniosis with aminosidine at an optimized dosage regimen: a pilot open clinical trial Veterinary Parasitology. 2013; 192:91–7.

Antimicrobial efficacy gaseous ozone on berries and baby leaf vegetables

S. de Candia[1,*]**,T. Yaseen**[2]**, A. Monteverde**[1]**, C. Carboni**[3] **and F. Baruzzi**[1]

[1]Institute of Sciences of Food Production, National Research Council of Italy, V. G. Amendola 122/O, 70126 Bari, Italy
[2]CIHEAM/Mediterranean Agronomic Institute of Bari, Via Ceglie, 9, 70010 Valenzano (BA), Italy
[3]De Nora NEXT-Industrie De Nora S.p.A. Via Bistolfi, 35- 20134 Milan, Italy
*Corresponding author: e-mail: silvia.decandia@ispa.cnr.it, Phone: +39 080.5929384

Ozone, the triatomic form of oxygen, is a strong broad-spectrum antimicrobial agent widely used for improving food safety; it rapidly auto-decomposes to oxygen and does not leave residues. However, its antimicrobial efficacy against microorganisms contaminating foods is greatly reduced as it promptly reacts with food organic matter, and consequently, its concentration decreases. In the last years, following the consumers' demand for RTE foods, fruits and vegetables are usually manipulated, processed and cold stored for some days before being ready for consumption. Thus, in comparison with fresh fruits and vegetables, RTE produces lead to a new matter of safety concerns with a greater frequency of foodborne illnesses.

Aim of this work was to evaluate the impact of gaseous ozone treatments on microorganisms contaminating berries and baby leaf vegetables. In the case of berries, ozone was applied at 2000 nL L^{-1} for 5 min, or with continuous fumigation at 300 nL L^{-1} evaluating the effect on yeast and mold population, the microflora mainly responsible for fruit decay, during seven days of cold storage. As concerns baby leaf vegetables, ozoneation was constant (at 500nL L-1) for seven days but maintaining leaves at 4°C and 10°C, evaluating its effect on *Pseudomonadaceae*, bacterial population responsible for browning of leaves, as well as on total mesophilyc bacteria.

Ozone caused a significant reduction of fungal contaminants on treated berries, during the conservation as compared with untreated fruits, in both application mode 2000 nL L^{-1} for 5min or in continuous fumigation at 300 nL L^{-1}. The storage of baby leaves under 500 nL L^{-1}of ozone did not significantly affect the concentration of bacterial cells in any of the storage times evaluated at both 4°C and 10°C.

In conclusion, this study underlines as gaseous ozone can improve microbial quality of fruits and vegetables but the impact of its efficacy depends on both target microflora and treated vegetable. In the light of these results, the control of undesired microorganisms contaminating fruits and vegetables needs to be evaluated case by case

Keywords: food safety, RTE vegetables, shelf-life, ozone sanification, risk management

1. Introduction

Fresh-cut fruits and vegetables have been recently traced as responsible for human outbreaks depending by low quality of water used for washing and chilling the produce after harvest [1].

Among several new phisico-chemical disinfectant methods today available, ozone treatments have been recently evaluated useful in improving microbial safety of vegetables and water bodies used in processing plants.

In order to monitor *E. coli* O157:H7 on green peppers with ozone gas (from 2 to 8 mg / L) [2] demonstrated, in addition to ozone concentration and the contact time, that antimicrobial efficacy was directly influenced by relative humidity. The efficacy of inactivate *E. coli* O157:H7 was also demonstrated on lettuce and baby-carrots in which gaseous ozone reduced viable cell count up to 2,6 log cfu/ml depending on ozone concentration (7.6 mg/L) time of exposure (15 min) and relative humidity (80%) [3].

Different results were obtained when natural microbial population were considered as a marker of antimicrobial efficacy of ozone treatment instead of a single target microorganism. Ozone treatment (at 5000 mg L^{-1}) was found ineffective to reduce total mesophilic bacteria on strawberries [4]; results comparable with these are also those obtained for fresh-cut papaya [5]for which reduction of both mesophilic bacteria and coliforms resulted appreciable only after 30 min at 9 ppm ozone. Even though fungi are usually considered a more sensitive microflora compared to bacteria, scientific bibliography shows conflicting results.

Figs [6] or date palm fruits [7] reduced significantly bacteria, yeast and mould populations only after ozone treatment at least 5 ppm concentration.

However, antimicrobial efficacy of ozone treatment is often associated with chlorophyll degradation, loss of natural fruit color and increase in respiration rate [8,9,10,11].

The consumption of baby leaf vegetables as well as berries is encouraged due to the health benefits that derive from their high content in fibers, vitamins and polyphenols. However, both these fresh vegetables can spoil during refrigerate storage. About 36% of ready to eat vegetables spoilage are due to soft rot developed by

Erwinia e *Pseudomonas* [12] as well as 100% of berries can be contaminated by several spoiling moulds (eg. *Botrytis cinerea, Rhizopus, Alternaria, Penicillium* etc.) [13].

Aim of this work was to evaluate the suitability of gaseous ozone treatments in controlling spoilage microorganisms (fungi and bacteria) contaminating ready-to-eat vegetables (berries and baby leaf).

2. Material and Methods

2.1 Experimental set up

Baby leaf samples homogeneous for type, weight and quality, were packed in trays of polypropylene and stored up to 7 days in the refrigerator at two temperatures (4°C and 10°C) and under three gaseous ozone concentration (0, 0.5 and 2 ppm). Raspberries, strawberries and blueberries used in this study were harvested from commercial fields located in Pergine Valsugana (Trento). Berries, packed in trays of polypropylene, were stored at 4°C for 7 days under ozone (300 nL L^{-1}) or after a single gaseous ozone treatment (2000 nL L^{-1} for 5min); untreated refrigerated berries were analysed as negative control. Ozone generator model MWP2.5, was provided by De Nora NEXT, Milan.

2.2 Microbiological analyses

At 0, 3 and 7 days 20 grams of leaves were homogenized in 180ml of sterile saline solution using Stomacher® 80 Biomaster (Seward Limited, West Sussex, UK), decimally diluted and plated onto PCA and PSA (Biolife Italiana srl, Milan, Italy) Petri dishes, for total mesophilic aerobic bacteria and presumptive pseudomonads, respectively. After two days of incubation at 30°C, microbial viable count was calculated by enumerating colonies developed onto each Petri dish and multiplying this value for dilution factor.

To quantify colony-forming units from treated and untreated berries, 25g of berries were subjected to orbital shaking at 150 rpm for 1hr in presence of 225 mL of sterilized distilled water. Decimally diluted samples were seeded onto NYDA medium [14]; colonies, enumerated after 3 days of incubation at 24°C, were then converted in CFU/g as described above.

2.3 Evaluation of antioxidant enzymes activity and polyphenols concentration

Superoxide dismutase (SOD), catalase (CAT) and glutathione peroxidase (GPX) were extracted as previously described [15], Enzymatic activities were analyzed spectrophotometrically [16] by using a Beckman DU 530 spectrophotometer. Anthocyanins and flavonols were extracted from lyophilized fruit samples and analyzed by HPLC following the method described by Nicoletti et al. [17].

2.4 Statistical analyses

The data were statistically evaluated using one-way analysis of variance (ANOVA) and Tukey post-hoc test (SPSS software, IBM Corporation, Armonk, NY, USA). Each value was reported as means ± standard deviation.

3. Results

3.1 Assessment of microbial growth

Baby leaf samples stored under 2 ppm ozone lost their fresh green appearance already after 3 days of cold storage at both temperatures. Since these samples showed unacceptable visual quality, they were not subjected to any microbiological analysis.

As concerns un-ozoneated baby leaf samples, total mesophilic aerobic bacteria load increased from 6.06 ± 0.35 log cfu/g to 7.87 ± 0.12 log cfu/g after 7 days of cold storage at both temperatures. In a similar way, presumptive pseudomonas, increased from 5.80 ± 0.43 log cfu/g to 8.14 ± 0.8 log cfu/g. Baby leaves weekly subjected to 0.5 ppm of gaseous ozone showed both total mesophilic aerobic bacteria and pseudomonads loads were not significantly (P > 0.05) lower than those of control samples.

These results are in agreement with those of other researchers [4,5] that found no or low reduction in natural contaminant bacterial population of fruit and vegetable after ozone treatment. In particular a slight reduction in total mesophilic bacteria and coliform populations was found by Yeoh et al. [5] but only after 30 min of ozone treatment at 9ppm, a concentration 4.5 times higher than that we found to waste baby leaves. Part of this resistance can come from ability of bacteria to be internalized into stomata where they are protected by leaf tissues from both hypochlorite solutions and ozonated water [18].

As concerns berry fruits ozone treatment, ozone caused a different yeasts and moulds reduction depending on berries type and initial contamination level. In particular, after a day of cold storage, blueberries and raspberries yeast and mould viable population was reduced from 2.30 ± 0.13 and 1.91 ± 0.20 log cfu/g to about 1.72 ± 0.24 log cfu/g independently by the ozone treatment applied; fungal population of 1.84 ± 0.29 log cfu/g was found in the un-ozoneated berries after the same storage period.

Extension of cold storage caused a slight increase in viable yeast and mould populations in both berries and for all treated and un-treated samples. Interestingly, this microbial population increased more than that of ozoneated samples even though in a not significant manner (P > 0.05), suggesting a possible tail of the ozone treatment on cell viability.

Ozone treatments produced a different microbial death kinetic in controlling yeasts and moulds naturally occurring on strawberries (Fig. 1).

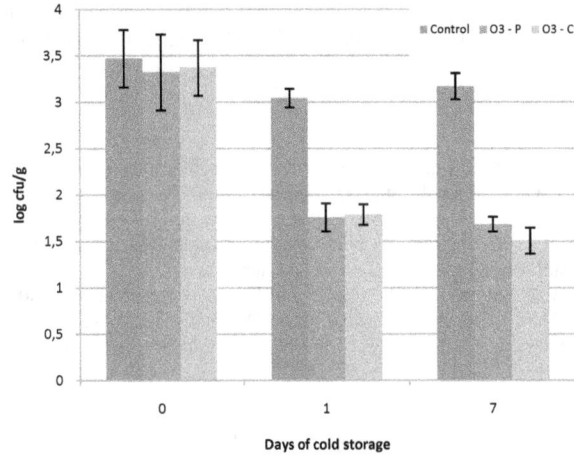

Fig. 1. Total yeast and mould population on strawberries after 7 days of cold incubation at ozone concentration 2000 nL L^{-1} for 5min (O3 – P) or in continuous fumigation at 300 nL L^{-1} for 7 days (O3 – C).

The efficacy that we found in treating with ozone high contaminated fruits, confirms previously published results on date fruits [7] and dried figs [6] even though these results were obtained with ozone concentration up to 5 ppm.

3.2 Changes in enzymatic activities and polyphenol content

As ozone may be involved in plant oxidative stress response, superoxide dismutase (SOD), catalase (CAT) and glutathione peroxidase (GPX) activities were measured in order to evaluate berries response to ozone oxidative stress; for the same reason the changes in antioxidant polyphenol concentration was recorded. Ozone treatment caused a significant increase (P < 0.05) in CAT activity but little or no change in SOD and GPX (Fig. 2). CAT activity usually rises in correlation with the increase of SOD activity, since catalase detoxifies H_2O_2 produced by SOD, but some works report that, following a mild stress, CAT activity can significantly increase even if SOD remains unchanged [19]. Following exposure of fruits to ozone, anthocyanins and flavonols, endowed with a strong antioxidant action, significantly decreased (P < 0.05) (Fig. 3). The reduction in their concentration could come from their scavenging action against free radical species, such as hydroperoxyl, hydroxyl, and superoxide radicals that can be produced by reaction with ozone [20].

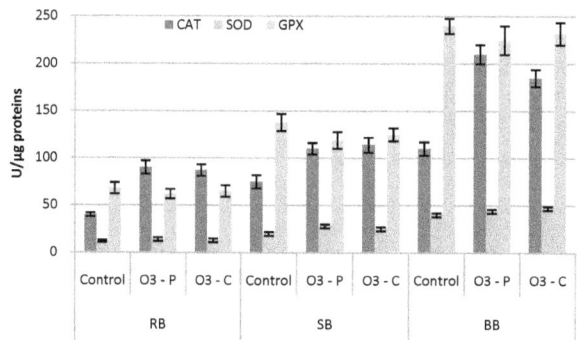

Fig. 2. Enzymatic activities of superoxide dismutase (SOD), catalase (CAT) and glutathione peroxidase (GPX) calculated for raspberries (RB), strawberries (SB) and blueberries (BB) stored at 4°C for 7 days. Samples are indicated as Control, O3-P (treated with ozone at 2000 nL L^{-1} for 5min) and O3-C (continuously 300 nL L^{-1} for 7 days).

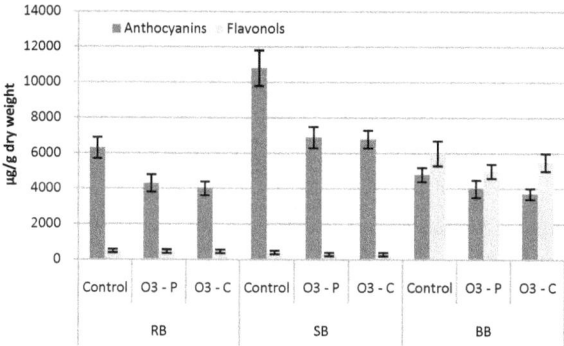

Fig. 3. Concentration of anthocyanins and flavonols for raspberries (RB), strawberries (SB) and blueberries (BB) stored at 4°C for 7 days. Samples are indicated as Control, O3-P (treated with ozone at 2000 nL L^{-1} for 5min) and O3-C (continuously 300 nL L^{-1} for 7 days).

This work shows that ozone treatments cause different kind of response in vegetable tissues, as depigmentation and dryness as well as changes in the concentration of main polyphenols. We can conclude that in addition to different efficacy in controlling different microbial spoiling microflora, the extent of the responses caused by ozone treatment depend on various factors, including the kind of vegetable tissue, as well as its colonization by microorganisms.

In conclusion, this study underlines as gaseous ozone can improve microbial quality of fruits and vegetables but the impact of its efficacy depends on both target microflora and treated vegetable. In the light of these results, the control of undesired microorganisms contaminating fruits and vegetables needs to be evaluated case by case

References

[1] Gil MI, Selma MV, López-Gálvez F, Allende A. Fresh-cut product sanitation and wash water disinfection: Problems and solutions. International Journal of Food Microbiology. 2009; 134:37-45.

[2] Han Y, Floros JD, Linton RH, Nielsen SS, Nelson PE. Food Microbiology and Safety Response Surface Modeling for the Inactivation of *Escherichia coli* O157:H7 on Green Peppers (*Capsicum annuum*) by Ozone Gas Treatment. Journal of Food Science. 2002; 67:1188-1193.

[3] Singh N, Singh RK, Bhunia AK, Stroshine RL. Efficacy of Chlorine Dioxide, Ozone, and Thyme Essential Oil or a Sequential Washing in Killing *Escherichia coli* O157: H7 on Lettuce and Baby Carrots. LWT-Food Science and Technology. 2002; 35:720-729.

[4] Allende A Mar′ın, Buend′ıa B, Tom′as-Barber′an F, Gil MI. Impact of combined postharvest treatments (UV-C light, gaseous O3, superatmospheric O2 and high CO2) on health promoting compounds and shelf-life of strawberries. Postharvest Biology and Technology. 2007; 46:201-211.

[5] Yeoh WK, Ali A, Forney CF. Effects of ozone on major antioxidants and microbial populations of fresh-cut papaya. Postharvest Biology and Technology. 2014; 89:56-58.

[6] Öztekin S, Zorlugenç B, Zorlugenç FK. Effects of ozone treatment on microflora of dried figs. Journal of food engineering. 2006; 75:396-399.

[7] Habibi N, Mohammad B, Haddad Khodaparast MH. Efficacy of ozone to reduce microbial populations in date fruits. Food Control. 2009; 20:27-30.

[8] Aguayo E, Escalona VH, Artes F. Effect of cyclic exposure to ozone gas on physicochemical, sensorial and microbial quality of whole and sliced tomatoes. Postharvest Biology and Technology. 2006; 39:169-177.

[9] Gabler FM, Smilanick JL, Mansour MF, Karaca H. Influence of fumigation with high concentrations of ozone gas on postharvest gray mold and fungicide residues on table grapes. Postharvest Biology and Technology. 2010; 55:85-90.

[10] Goncalves AA. Ozone—an emerging technology for the seafood industry. Brazilian Archives of Biology and Technology. 2009; 52:1527-1539.

[11] Wang YJ, Pan MH, Cheng AL, Lin LI, Ho YS, Hsieh CY, Lin JK. Stability of curcumin in buffer solutions and characterization of its degradation products. Journal of Pharmaceutical and Biomedical Analysis. 1997; 15:1867-1876.

[12] Sudheer KP, Indira V. Post harvest technology of horticultural crops. New India Pub. Agency, Pitam Pura, New Delhi; 2007.

[13] Tournas VH, Katsoudas, E. Mould and yeast flora in fresh berries, grapes and citrus fruits. International Journal of Food Microbiology. 2005; 105:11-17.

[14] Ricelli A, Baruzzi F, Solfrizzo M, Morea M, Fanizzi FP. Biotransformation of patulin by *Gluconobacter oxydans*. Applied and Environmental Microbiology. 2007; 73:785-792.

[15] Reverberi M, Fabbri AA, Zjalic S, Ricelli A, Punelli F, Fanelli C. Antioxidant enzymes stimulation in *Aspergillus parasiticus* by *Lentinula edodes* inhibits aflatoxin production. Applied Microbiology and Biotechnology. 2005; 69:207-215.

[16] Kim JH, Campbell BC, Yu J, Mahoney N, Chan KL, Molyneux RJ, Bhatnagar D, Cleveland E. Examination of fungal stress response genes using *S. cerevisiae* as a model system: targeting genes affecting aflatoxin biosynthesis by *Aspergillus flavus*. Applied Microbiology and Biotechnology. 2005; 67:807-815.

[17] Nicoletti I, Bello C, De Rossi A, Corradini D. Identification and quantification of phenolic compounds in grapes by HPLC-PDA-ESI-MS on a semimicro separation scale. Journal of Agricultural and Food Chemistry. 2008; 56:8801-8808.

[18] Saldaña Z, Sánchez E, Xicohtencatl-Cortes J, Puente JL, Girón JA. Surface structures involved in plant stomata and leaf colonization by Shiga-toxigenic *Escherichia coli* O157: H7. Frontiers in Microbiology. 2011; 2:119.

[19] Yongqing T, Potempa J, Pike RN, Wijeyewickrema LC. The lysine-specific gingipain of *Porphyromonas gingivalis*: importance to pathogenicity and potential strategies for inhibition. Advances in Experimental Medicine and Biology. 2011; 712:15-29.

[20] Alothman M, Kaur B, Fazilah A, Bhat R, Karim AA. Ozone-induced changes of antioxidant capacity of fresh-cut tropical fruits. Innovative Food Science & Emerging Technology. 2010; 11:666-671.

Antimicrobial hop extracts and their application on fresh produce

C. Hauser[*,1] and **T. Sentürk Parreidt**[1]

[1] Fraunhofer Institute for Process Engineering and Packaging IVV, Giggenhauser Straße 35, 85354 Freising, Germany

*Corresponding author: e-mail: carolin.hauser@ivv.fraunhofer.de, Phone: +49 8161491626

Fresh-cut fruits and vegetables get a growing share on the worldwide food market. Nevertheless, these non-treated food products bear a risk of microbial contamination with pathogenic microorganisms and limited shelf-life due to their fresh character and big surface. Treating food with antimicrobial substances can minimize the inherent microbiological risk and is even able to prolong shelf-life and increase the quality of these products. On the other hand chemically synthesized preservatives get more and more rejected by the consumer. Natural substances are preferred instead. For example lupulone, a beta-acid of the hop plant, has a high antimicrobial potential.

These extracts have been applied on fresh-cut produce such as endive salad and cantaloupe melons. The application was either directly or via an edible coating. The efficiency of the extracts on the quality of the products was evaluated in storage tests. They showed that natural hop extracts were able to contribute significantly to the safety and quality of fresh produce. Thus they could be a serious alternative to conventional additives for food preservation.

Keywords hop; beta-acids; natural antimicrobials; food preservation, fresh-cut; edible coating

1. Introduction

Fresh-cut fruits and vegetables combine the aspect of convenience food with healthy nutrition. Therefore, they get a growing share on the worldwide food market as a result of changes in consumer attitudes, i.e. increasing demand of fresh, healthy and convenient food.

Fresh-cut technology implies peeling, slicing, dicing or shredding prior to packaging and storage. These processes accelerate physiological deterioration, biochemical changes and microbial degradation of the product [1, 2, 3]. Microbial activity increases as a consequence of increased disruption of surface cells and neutral pH. With a high surface/weight ratio, fresh-cut products host a large microbial population, particularly bacteria [2]. Additionally, these non-treated food products bear a risk of microbial contamination with pathogenic microorganisms. Concomitant with an increased consumption of these products, increased numbers of foodborne outbreaks have been associated with fresh produce [4].

Retailing in cold chain regime and the modified atmosphere applications are weak hurdles. Therefore defining new, natural food preservatives against microorganisms will not only meet the antimicrobial preservation but also retaining nutritional and sensory quality for consumers. Beta acids containing natural hop extracts exert high bactericidal effect on various food-related (pathogenic) bacteria. Especially gram-positive bacteria such as Listeria are very sensitive to hop beta-acids [5,6].

The objective of the study was to reduce the total viable count on fresh-cut products such as salad and cantaloupe melon, with the application of beta-acids containing extracts. The application was by dipping the products either directly in an aqueous solution containing beta-acids (salad) or via an edible coating based on alginate with beta-acids (melons).

The efficiency of the extracts on the total viable count of the products was evaluated in storage tests.

2. Materials and Methods

In order to evaluate the antimicrobial efficiency of the hop extracts on fresh-cut products they were applied in two different ways:

2.1 Direct application on fresh-cut salad

Fresh endive salad *(Cichorium endivia)* was purchased from a local market, washed with cold water and cut into pieces of about 5 cm². Subsequently the pieces were dipped for 45 s into an aqueous solution containing 1% citric acid and 0.05% beta-acids containing hop-extract (BetaBio 40, Hopsteiner, Germany). As a reference, pieces of salad were dipped in the same way in 1% citric acid solution. Excessive water was shaken off manually. The leaves were dried for 30 minutes and placed in sterile polyethylene bags.

The fresh-cut products were stored in darkness at 8 °C for up to 8 days in an environmental chamber. After several time intervals a total viable count of at least five independent samples was performed.

3 g of salad was homogenized in Tween 80 added Ringer's solution (97 ml) in a stomacher (Smasher, AES Chemunex, Canada) for 2 minutes at 260 rpm. Decimal serial dilutions were prepared subsequently using 9 ml volumes of ringer solution. Appropriate dilutions were pour-plated with sterile plate count agar (Merck kGaA, Germany) and incubated at 37 °C for 48 hours.

2.2   Application via edible coating on fresh-cut cantaloupe melon

Cantaloupe melons (*Cucumis melo* L. var. *reticulatus group*) used in this study were purchased from a local market where they had been stored at ambient temperature. All cantaloupe melons were used one day after purchase. They were peeled and cut into pieces of about 8 g. Coating solution was prepared by dissolving 1.25 g of sodium alginate (FMC Cooperation, USA) in 90 g sterilized distilled water while stirring on a magnetic stirrer/hot plate (Heidolph MR 3001, Germany) at 70°C until the solution became clear. 2 g glycerol (Sigma-Aldrich Co., USA) was added to increase plasticity of the coating and was mixed into the solution without heating for 15 minutes, until it was dispersed. 0.125 g sunflower oil was added with the following composition: 10.1 g saturated fatty acids, 25.7 g mono and 56.0 g polyunsaturated fatty acids in 100 ml oil, in order to improve water vapour barrier properties [7]. Polyoxyethylenesorbitan monopalmitate (Tween 40) (Sigma-Aldrich Co., USA) and sorbitane monooleate (Span 80) (Sigma-Aldrich Co., USA) were added as surfactant at 0.1 g and 0.6 g, respectively. 62.5 mg hop extract (BetaBio40, Hopsteiner, Germany) was added as antimicrobial solution. Weight of the solution was increased to 100 g with addition of water. The solution was homogenized for 5 minutes at 10,000 rpm using Ultra Turrax *Art Micra D8* (*Micra* RT Labortechnik, Germany).

In order to induce the cross-linking reaction, 2 g calcium lactate (Merck KGaA, Germany) was dispersed in 90 g sterilized distilled water. 62.5 mg hop extract was added as antimicrobial agent. The weight of the solution was increased to 100 g with addition of water.

After each dipping process, pieces were allowed to drain for about 2 minutes. As a reference, pieces of melon without any coating treatment were used. About 150 g of the fruit pieces were transferred into EPS Trays and closed with stretch films.

The fresh-cut products were stored in darkness at 8 °C for up to 8 days in an environmental chamber. After several time intervals a total viable count of at least three independent samples was performed.

30 g of cantaloupe was homogenized in Tween 80 added ringer solution (270 ml) in a stomacher (Smasher, AES Chemunex, Canada) for 2 minutes. Decimal serial dilutions were prepared subsequently using 9 ml volumes of ringer solution. Appropriate dilutions were pour-plated with sterile plate count agar (Merck kGaA, Germany) and incubated at 37 °C for 48 hours.

3. Results and Discussion

3.1   Direct application

The development of the total viable count on fresh-cut endive salad is shown in Figure 1. The colony count on the untreated endive salad could be reduced due to the washing/dipping step in citric acid by about 2 (reference) to almost 3 log-cycles (hop-extract). Citric acid is a weak organic acid, often applied in food industry. In aqueous solution it is partly undissociated and thus can pass the cell membrane and interact adversely in the cell metabolism. A mean difference of at least 0.5 log cycles could be observed throughout the storage time between the hop-extract containing samples and the references. On the seventh day of storage the total viable count was even reduced to 1 log cycle compared to the reference sample and 2 log cycles compared to the initial count. As the efficacy of hop beta-acids can be enhanced by the addition of weak organic acids and thus lowering the pH [8], citric acid had a positive effect on the antimicrobial activity of the hop-containing washing solution.

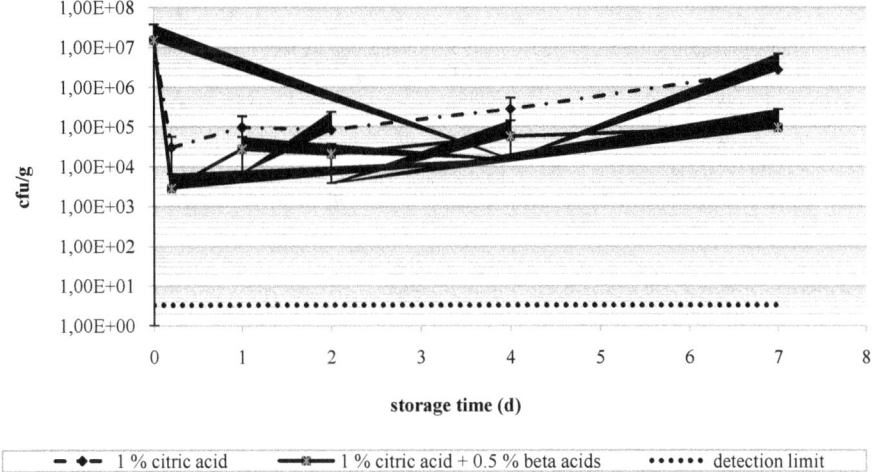

Fig. 1 Development of the total viable count on fresh-cut salad after dipping (45 s) in 1% citric acid and 1% citric acid with addition of beta-acid containing hop-extract (0.05%) over 7 days packed in polyethylene films; storage temperature: 8 °C; data are expressed as means ± SD, n=5 [9]

3.2 Application via edible coating

The result of initial total count of cantaloupe samples agreed with the literature [10, 11]. Incorporated into the edible coating the hop extract showed bacteriostatic effect during the first 24 h on the cantaloupe melons (Figure 2). Compared to the reference, the count reduction increased from 1 log cycle after the first day to 2 log cycles after the fourth day. After reaching about 10^6 cfu/g at the sixth day there was no significant difference observable at the eighth day. This can be due to the evolution of yeasts, which cannot be inhibited by the beta-acids [12]. The maximum microbial limit fixed by German Society for Hygiene and Microbiology (DGHM) [13] for aerobic growth (10^7 cfu/g) was nearly reached after the sixth day and exceeded after 7 days (control group) and 8 days (coated group) of storage at 8°C. Thus the shelf-life could be extended by at least one day.

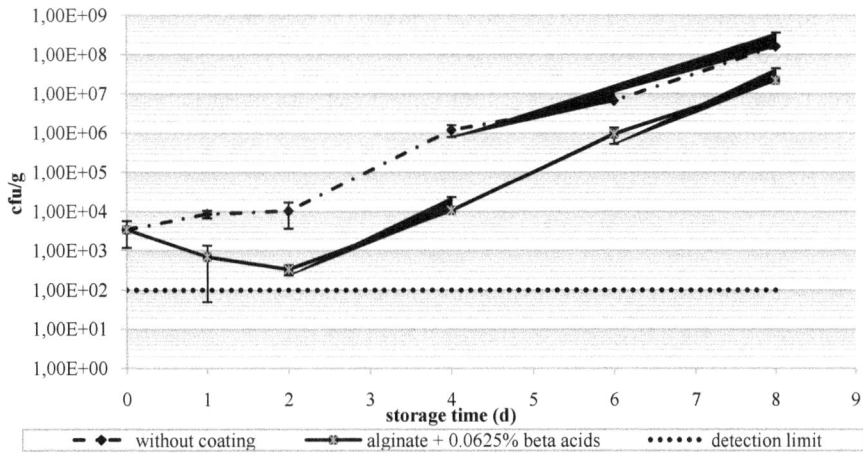

Fig. 2 Development of the total viable count on fresh-cut cantaloupe melon after dipping in (2 min) alginate and (2 min) Ca-lactate solution with addition of beta-acid containing hop-extract (0.0625%) over 8 days packed in expanded polystyrene (EPS) trays; storage temperature: 8 °C; data are expressed as means ± SD, n=3 [8]

4. Conclusions

Citric acid solution and alginate based edible coating containing hop extract were prepared for dipping process of endive salad and cantaloupe melon, respectively. For keeping microbial safety of fresh-cut products, hop extract was added to dipping and edible coating solution. It is concluded that beta-acids containing hop extracts had a positive effect on the evolution of bacterial growth on both fresh-cut produce. As beta-acids mainly inhibit gram-positive bacteria, the effect strongly depends on the food product and its initial microflora. Incorporating hop extracts into edible coatings seems to favor a slow but steady release of the antimicrobial substances to the food product until a certain threshold is reached. Thus they could be a serious alternative to conventional additives for food preservation.

Acknowledgements This research was carried out in two projects (16975 N and 95 EN/1) under the auspices of the German Federation of Industrial Research Associations (AiF). It is financed within the budget of the German Federal Ministry of Economics and Technology (BMWi) through the program to promote collective industrial research (IGF). The research project IGF-Nr. 95 EN/1 was implemented under the CORNET initiative.

References

[1]   Rico D, Martin-Diana AB, Barat JM, Barry-Ryan C. Extending and measuring the quality of fresh-cut fruit and vegetables: a review. Trends in Food Science & Technology. 2007; 18:373-86.

[2]   Caponigro V, Ventura M, Chiancone I, Amato L, Parente E, Piro F. Variation of microbial load and visual quality of ready-to-eat salads by vegetable type, season, processor and retailer. Food Microbiology. 2010; 27:1071-77.

[3]   Portela SI, Cantwell MI. Cutting blade sharpness affects appearance and other quality attributes of fresh-cut cantaloupe melon. Journal of Food Science. 2001; 66(9):1265-70

[4]   Doyle MP, Erickson MC. Review article summer meeting 2007 – the problems with fresh produce: an overview. The Society for Applied Microbiology, Journal of Applied Microbiology. 2008; 105:317–30.

[5]   Shen C, Sofos JN. Antilisterial activity of hops beta acids in broth with or without other antimicrobials. Journal of Food Science. 2008; 73(9): M438-M442

[6]   Teuber M, Schmalreck AF. Membrane leakage in *Bacillus subtilis* 168 induced by the hop constituents lupulone, humulone, isohumulone and humulinic acid. Archives of Microbiology. 1973; 94(2):159-71.

[7]   Rojas-Graü MA, Tapia MS, Rodríguez FJ, Carmona AJ, Martin-Belloso O. Alginate and gellan-based edible coatings as carriers of antibrowning agents applied on fresh-cut Fuji apples. Food Hydrocolloids. 2007; 21(1):118-27.

[8]   Simpson, W., Smith, A. Factors affecting antibacterial activity of hop comounds and their derivates. Journal of Applied Microbiology. 1992, 72(4): 327-334

[9]   Fraunhofer IVV (2014) internal results

[10] Aguayo E, Allende A, Artés F. Keeping quality and safety of minimally fresh processed melon. European Food Research and Technology. 2003; 216:494-9.

[11] Bai JH, Saftner RA, Watada AE, Lee YS. Modified atmosphere maintains quality of fresh-cut cantaloupe (*Cucumis melo* L.). Journal of Food Science. 2001; 66:1207-11.

[12] Srinivasan V, Goldberg D, Haas GJ. Contributions to the antimicrobial spectrum of hop constituents. Economic botany 2004; 58(1):S230-S238.

[13] German Society for Hygiene and Microbiology (DGHM), Mikrobiologische Richt- und Warnwerte zur Beurteilung von Lebensmitteln, 2012

Antimicrobial resistance of *Escherichia coli* isolated from small animals in Lithuania: a cross sectional study

M. Ruzauskas, M. Virgailis[*], R. Siugzdiniene, I. Klimiene, L. Vaskeviciute, S. Ramonaite, J. Zymantiene and R. Mockeliunas

Veterinary Academy, Lithuanian University of Health Sciences, Tilzes g. 18, Kaunas, Lithuania
*Corresponding author: e-mail: marius.virgailis@lsmuni.lt, Phone: 00370 61004014

The aim of this study was to isolate *E. coli* from small animals and to determine resistance to antimicrobial classes important for humans. In 2012-2014 three hundred and eighty animals (272 dogs, 55 cats and 53 other animals) were tested for the presence of *E. coli*. Both, diseased (n=280) and healthy (n=100) animals, were included in this study.

Two hundred and sixty isolates of *E. coli* were obtained (68%) from the animals tested. The most common resistances were demonstrated to ampicillin (39%), sulfamethoxazole/trimethoprim (16%) and ciprofloxacin (10%). Resistance to gentamicin was detected in 5% of the isolates. Extended spectrum beta-lactamases were produced by 3.1% of the isolates with attribution to the *tem* gene. The only one isolate harboured the CTX-M gene. Other genes encoding resistances included *sul1*, *sul2* and *sul3* (sulphonamides), *dfr1*, *dfr5* and *dfrA7* (trimethoprim), *aac(3)II*, *aphA1* and *aadA* (aminoglycosides).

Keywords genes; antimicrobial resistance; small animals; *E. coli*

1. Introduction

Escherichia coli (*E. coli*) is a common inhabitant of the gastrointestinal tract of humans and animals. Usually, this species forms a beneficial symbiotic relationship with its host and plays important roles in promoting the stability of the luminal microbiota and in maintaining normal intestinal homeostasis [1]. As a commensal, *E. coli* rather remains harmlessly confined to the intestinal lumen and rarely causes a disease. However, in the debilitated or immunosuppressed host, or when the gastrointestinal barriers are violated, even nonpathogenic-commensal strains of *E. coli* can cause infection [2]. Some strains of *E. coli* can diverge from their commensal cohorts, taking on a more pathogenic nature. These strains acquire specific virulence factors (via DNA horizontal transfer of transposons, plasmids, bacteriophages, and pathogenicity islands), which confer an increased ability to adapt to new niches and allow the bacteria to increase the ability to cause a broad spectrum of diseases [3]. Animals, including dogs are known as a possible reservoir of the virulent *Escherichia coli* strains that cause extra intestinal infections in humans [4]. The aim of this study was to isolate *E. coli* from small animals and to determine resistance to antimicrobial classes important for humans.

2. Materials and methods

2.1 Study design, animals and sampling

In 2012-2014 clinical samples were collected from 380 pet animals on small animal clinics in Lithuania. The samples were taken from both diseased (n=280) and healthy animals (n=100). Diseased individuals included animals with infections of gastrointestinal and respiratory tract, reproductive organs and urinary tract. Samples were collected from rectum by veterinary surgeon using sterile Amies media swabs (Liofilchem, Italy). Samples were delivered to the laboratory during the same day. This study involved animals from the five counties of Lithuania.

2.2 Bacteriological analyses

Clinical material was inoculated onto MacConkey Agar (Liofilchem, Italy), Tryptone Bile X-Glucuronide (*TBX*) Agar (Biolife, Italy) and Colorex ESBL Agar (E&O Laboratories, Scotland).

Presumptive identification of *E. coli* was based on the growth and morphology characteristics and indole production. Microgen GN-ID system (Microgen, UK) was used for species identification according to biochemical testing. Software MID (Version 1.2, Microgen) was used for the interpretation of results. In uncertain cases, 16S rRNA sequencing for species confirmation was performed using an ABI3730XL sequencer. The universal primers 27F and 515R were used as described previously [5]. Sequences were analysed using Molecular Evolutionary Genetic Analysis software (MEGA, version 6). Basic local alignment search tool (BLAST) was used for comparison of obtained sequences with sequences presented in the database of National centre of biotechnology information.

2.3 Molecular testing of genes encoding resistance

DNA material for molecular testing was obtained after bacterial lysis according to the extraction protocol prepared by the Community Reference Laboratory for Antimicrobial Resistance with slight modifications. Briefly, a loopful of colonies was taken from the surface of Mueller Hinton Agar and transferred to phosphate buffered saline (pH 7.3). The content was centrifuged for 5 min. Then the supernatant was discarded and the pellet was re-suspended in Tris-EDTA (TE) buffer. The suspension was heated using a thermomixer at 100 °C degrees for 10 minutes. Boiled suspension was transferred directly on ice and diluted by 1:10 in TE. PCR was used for the detection of genes encoding resistance to beta-lactams *(tem, CTX-M,)*, tetracyclines (*tetA, tetB*), aminoglycosides (*strA, strB, aadA, aphA, aac(3)II*), trimethoprim (dfrA1, dfrA5,dfrA7) and sulphonamides (*sul1, sul2, sul3*) (Table 1).

Table 1 Oligonucleotide primers used in this study

Primer name	Sequence(5'-3')	Size, bp and t(°C)	Target gene	Source
*bla*TEM-F	GAGTATTCAACATTTTCGT	857 (50)	*tem*	Van et al., 2008
*bla*TEM-R	ACCAATGCTTAATCAGTGA			
CTX-MF	ATGTGCAGYACCAGTAARGT	593 (50)	*CTX-M*	Pagani et al., 2003
CTX-MR	TGGGTRAARTARGTSACCAGA			
sul1-F	TTCGGCATTCTGAATCTCAC	822 (55)	*sul1*	Christabel et al., 2012
sul1-R	ATGATCTAACCCTCGGTCTC			
sul2-F	CGGCATCGTCAACATAACC	722(50)	*sul2*	Sáenz et al., 2004
sul2-R	GTGTGCGGATGAAGTCAG			
sul3-F	GAGCAAGATTTTTGGAATCG	792(51)	*sul3*	Perreten et al., 2003
sul3-R	CATCTGCAGCTAACCTAGGGCTTTGGA			
Dfr1-F	ACGGATCCTGGCTGTTGGTTGGACGC	254(55)	*dfrA1*	Gibreel, Skold, 1998
Dfr1-R	CGGAATTCACCTTCCGGCTCGATGTC			
Dfr5-F	GCBAAAGGDGARCAGCT	394(44)	*dfrA5*	Šeputienė et al., 2010
Dfr5-R	TTTMCCAYATTTGATAGC			
DfrA7-F	AAAATTTCATTGATTTCTGCA	471(44)	*dfrA7*	Navia et al., 2003
DfrA-R	TTAGCCTTTTTTCCAAATCT			
tetA-F	GTGAAACCCAACATACCCC	887 (50)	*tet*(A)	Van et al., 2008
tetA-R	GAAGGCAAGCAGGATGTAG			
tet(B)-F	CCTTATCATGCCAGTCTTGC	773 (50)	*tet*(B)	Van et al., 2008
tet(B)-R	ACTGCCGTTTTTTCGCC			
strA-F	CCTGGTGATAACGGCAATTC	546 (55)	*str*(A)	P. Boerlin, 1999
strA-R	CCAATCGCAGATAGAAGGC			
strB-F	ATCGTCAAGGGATTGAAACC	509 (55)	*str*(B)	P. Boerlin, 1999
strB-R	GGATCGTAGAACATATTGGC			
aadA-F	GTGGATGGCGGCCTGAAGCC	526 (68)	*aadA*	Sandvang et al., 1997
aadA-R	AATGCCCAGTCGGCAGCG			
aphA1-F	AAACGTCTTGCTCGAGGC	461 (55)	*aphA1*	Frana et al., 2001
aphA1-R	CAAACCGTTATTCATTCGTGA			
aac(3)II-F	TGAAACGCTGACGGAGCCTC	369 (65)	*aac(3)*II	Sandvang, Aarestrup, 2000
aac(3)II-R	GTCGAACAG GTAGCACTGAG			

2.4 Antimicrobial susceptibility testing

Antimicrobial susceptibility testing was performed using the broth micro-dilution method. Sensititre GN2F plates and the ARIS 2X automated system (Thermo Scientific) were used with the following antimicrobials: amikacin, ampicillin, ampicillin/sulbactam, aztreonam, cefazolin, cefepime, cefotetan, ceftriaxone, ceftazidime, cefoxitin, cefuroxime, ciprofloxacin, gentamicin, imipenem, gatifloxacin, meropenem, piperacillin, nitrofurantoin, piperacillin/tazobactam, ticarcillin/clavulanic acid, tobramycin, sulfamethoxazole/trimethoprim, and cefpodoxime. Interpretation of results was carried out using manufacturer's software (SWIN®) adapted to clinical breakpoints of the European Committee on antimicrobial susceptibility testing (EUCAST). The quality control strain *E. coli* ATCC 25922 was used for validation purposes.

2.5 Statistical analysis

Statistical analysis was performed using "R 1.8.1" package (http://www.r-project.org/). Comparison between categorical variables was calculated by chi-square and Fisher's exact test. Results were considered statistically significant if P<0.05.

3. Results and discussion

Two hundred and sixty isolates (68%) of *E. coli* were obtained from the animals tested. The most common resistances were demonstrated to ampicillin (39%), sulfamethoxazole/trimethoprim (16%) and ciprofloxacin (10%) (Fig. 1). Resistance to gentamicin was detected in 5% of the isolates. *E. coli* isolated from dogs were more frequently resistant to fluoroquinolones, aminoglycosides and beta-lactams, while cat isolates statistically more often demonstrated resistance to sulfamethoxazole/trimethoprim (p≥0.05).

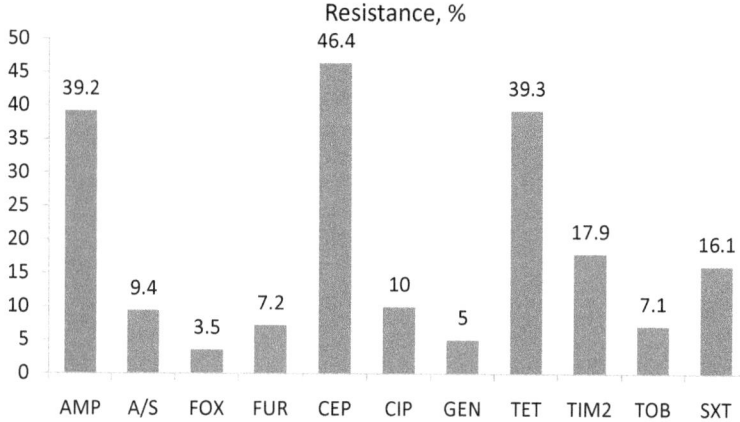

AMP-ampicillin, A/S-ampicillin-sulbactam, FOX-cefoxitin, FUR-cefuroxime, CEP-cephalothin, CIP-ciprofloxacin, GEN-gentamicin, TET-tetracycline, TIM2- ticarcillin/clavulanic acid, TOB-tobramycin, SXT – sulphamethoxazole./trimethoprim

Fig. 1 Number of resistant isolates (%) of *E. coli* isolated from small animals in Lithuania

Extended spectrum beta-lactamases were produced by 3% of the isolates with attribution to the *tem* gene. The only one isolate harboured the CTX-M gene. Other genes encoding resistances included *sul1*, *sul2* and *sul3* (sulphonamides), *dfr1*, *dfr5* and *dfrA7* (trimethoprim), *aac(3)II*, *aphA1* and *aadA* (aminoglycosides) (Fig.2).

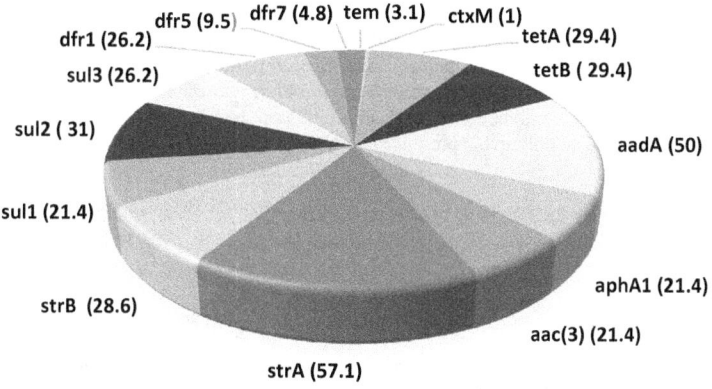

Fig. 2 Genes encoding resistance in *E. coli* isolated from small animals

The minimum inhibitory concentrations of antibiotics tested are presented in Table 2.

Table 2 MIC's of antimicrobials to *E. coli* isolates obtained from small animals (%), n=260

Antibiotics	MIC (mg/L)							
	0,5	1	2	4	8	16	32	64
Amikacin				100				
Ampicillin				47.2	13.6			39.2
Ampicillin-sulbactam				71.9	61.5	12.7	6.5	3.1
Aztreonam				100				
Cefazolin				89.3	7.1	3.6		
Cefepime				100				
Cefoxitin				64.6	28.4	3.5		3.5
Cefpodoxime			96.4	3.6				
Ceftazidime		96.9				3.1		
Ceftriaxone		96.9			3.1			
Cefuroxime					75	17.9	3.6	3.6
Cephalothin					46.4	46.4	7.1	
Ciprofloxacin	89.6		3.1	3.1	3.8			
Ertapenem			100					
Gentamicin			95.0		1.9		3.1	
Meropenem		100						
Piperacillin-tazobactam						100		
Tetracycline		35.7	25				39.2	
Ticarcillin-clavulanic acid						78.6	14.3	7.2
Tigecycline		100						
Tobramycin				82.1	10.7	7.1		
Co-trimoxazole	62.7		13.5	7.7	16.1			

As it could be seen from the data, the highest concentration of antibiotics for bacteria growth inhibition was detected for ampicillin and tetracycline (39.2 % of the isolates were inhibited by 64 mg/L or 32 mg/L of those antimicrobials respectively). Resistance to critically important classes of antimicrobials for humans was detected in some isolates. For example, 26 isolates were resistant to fluoroquinolones (ciprofloxacin) and 8 produced ESBL's by demonstrating resistance ceftazidime. *AmpC* type resistance to beta-lactams was detected in 9 isolates as well. Antimicrobials that are used exclusively in human medicine (i.e. amikacin, carbepenems, tigecycline) remain effective against all *E. coli* isolated from small animals in Lithuania. In some other countries the situation on antimicrobial resistance of *E. coli* isolated from domestic animals is less favourable. For instance, resistance to carbepenems of *E. coli* isolated from dogs was recently detected in the USA (Shaheen et al., 2013). Carbepenem-resistant isolates often are multi-resistant and infections caused by such bacteria are very hard to treat. Special attention should be paid to the possibility to improve the situation on antimicrobial resistance in pets: more strict and concrete indications should be approved, whereas off-label usage of antimicrobials should be applied only in the presence of severe infections caused by bacteria.

Acknowledgements This research was funded by a grant MIP-061/2012 from the Research Council of Lithuania.

References

[1] Yan F, Polk DB. Commensal bacteria in the gut: learning who our friends are. Current Opinion in Gastroenterology. 2004; 20:565-71.
[2] Kaper JB, Nataro JP, Mobley HLT. Pathogenic *Escherichia coli*. Nature Reviews Microbiology. 2004; 2:123-40.
[3] Bien J, Sokolova O, Bozko P. Role of uropathogenic *Escherichia coli* virulence factors in development of urinary tract infection and kidney damage. International Journal of Nephrology. 2012; Article ID 681473, 15 pages.

[4] Johnson JR, Stell AL, Delavari P. Canine feces as a reservoir of extraintestinal pathogenic *Escherichia coli. Infection and Immunology.* 2001; 69:1306-14.

[5] Kim JH, Lee JY, Kim HR, Heo KW, Park SK, Lee JN, Yu SM Shin JH. Acute lymphadenitis with cellulitis caused by *Staphylococcus lugdunensis.* Korean Laboratory Journal of medicine. 2008. 28: 196-200.

[6] Van TT, Chin J, Chapman T, Tran LT, Coloe PJ. Safety of raw meat and shellfish in Vietnam: an analysis of *Escherichia coli* isolations for antibiotic resistance and virulence genes. International Journal of Food Microbiology. 2008; 124:217-23.

[7] Pagani L, Dell'Amico E, Migliavacca R, D'Andrea MM, Giacobone E, Amicosante G, Romero E, Rossolini GM. Multiple CTX-M-type extended-spectrum-lactamases in nosocomial isolates of *Enterobacteriaceae* from a hospital in northern Italy. Journal of Clinical Microbiology. 2003; 41:4264-69.

[8] Christabel M, Budambula N, Kiiru J, Kariuki S. Full length research paper characterization of antibiotic resistance in environmental enteric pathogens from Kibera slum in Nairobi-Kenya. Journal of Bacteriology Research. 2012; 4:46-54.

[9] Sáenz Y, Briñas L, Domínguez E, Ruiz J, Zarazaga M, Vila J, Torres C. Mechanisms of resistance in multiple-antibiotic-resistant *Escherichia coli* strains of human, animal, and food origins. Antimicrobial Agents and Chemotherapy. 2004; 48:3996-01.

[10] Perreten V, Boerlin P. A new sulfonamide resistance gene (*sul3*) in *Escherichia coli* is widespread in the pig population of Switzerland. Antimicrobial Agents and Chemotherapy. 2003; 47:1169-72.

[11] Gibreel A, Skold O. High-level resistance to trimethoprim in clinical isolates of *Campylobacter jejuni* by acquisition of foreign genes (*dfr1* and *dfr9*) expressing drug-insensitive dihydrofolate reductases. *Antimicrobial Agents and Chemotherapy.* 1998; 42:3059-64.

[12] Šeputienė V, Povilonis J, Ružauskas M, Pavilonis A, Sužiedėliene E. Prevalence of trimethoprim resistance genes in *Escherichia coli* isolates of human and animal origin in Lithuania. Journal of Medical Microbiology. 2010; 59:315-22.

[13] Navia MM, Ruiz J, Sanchez-Cespedes J, Vila J. Detection of dihydrofolate reductase genes by PCR and RFLP. Diagnostic Microbiology and Infectious Disease. 2003; 46:295-98.

[14] Boerlin P. Associations between virulence factors of Shiga toxin-producing *Escherichia coli* and disease in humans. Journal of Clinical Microbiology. 1999; 37:497-03.

[15] Sandvang D, Aarestrup FM, Jensen LB. Characterisation of integrons and antibiotic resistance genes in Danish multiresistant *Salmonella enterica Typhimurium* DT104. FEMS Microbiology Letters. 1997; 157:177-81.

[16] Frana ATS, Carlson SA, Griffith RW. Relative distribution and conservation of genes encoding aminoglycoside-modifying enzymes in *Salmonella enterica* serotypeTyphimurium phage type DT104. Applied and Environmental Microbiology. 2001; 67:445-48.

[17] Sandvang D, Aarestrup FM. Characterization of aminoglycoside resistance and class 1 integrons in porcine and bovine gentamicin-resistant *Escherichia coli*. Microbial Drug Resistance. 2000; 6: 19-7.

[18] Shaheen BW, Navak R, Boothe DM. Emergence of a New Dehli metallo-β-lactamase (NDM-1)-encoding gene in clinical *Escherichia coli* isolates recovered from companion animals in the United States. Antimicrobial Agents and Chemotherapy. 2013; 57:2902-03.

Application of lactoferricin B to control microbial spoilage in cold stored fresh foods

L. Quintieri[1,*], A. Carito[1], L. Pinto[1], N. Calabrese[1], F. Baruzzi[1] and L. Caputo[1]

[1]Institute of Sciences of Food Production, National Research Council of Italy, V. G. Amendola 122/O, 70126 Bari, Italy

*Corresponding author: e-mail: laura.quintieri@ispa.cnr.it, Phone: +39 080.5929323

Psychrotrophic *Pseudomonas* spp. strains were proved to cause dramatic quality losses in fresh foods like leafy vegetables and dairy products resulting in the shortening of their shelf-life. These spoilage microorganisms are resistant to most disinfection treatments. This work was addressed to assay the antimicrobial efficacy of the food-grade bovine lactoferrin (BLF) and its hydrolysates (LFHs), obtained after digestion with pepsin or the enzymatic cardoon extract. The inhibition assays were carried out *in vitro* on selected *Pseudomonas* spp. strains involved in off-color development in ready-to-eat vegetables (RTE) and in mozzarella cheese.

A higher antimicrobial activity against most of the tested strains was found for pepsin digested BLF in comparison with BLF digested with the enzymatic cardoon extract. A solution of the antimicrobial peptide lactoferricin B (LFcinB), purified from the pepsin-LFH, was applied to both cold stored RTE lettuce leaves and mozzarella cheeses, inoculated with the spoilage strains. A significant reduced tissue browning of RTE lettuce leaves as well as the absence of mozzarella cheese discoloration were found applying LFcinB. The results obtained show the possibility to use milk-derived antimicrobial peptides in the control of food spoilage caused by *Pseudomonas* strains.

Keywords: *Pseudomonas* spp.; bovine lactoferrin; antimicrobial peptides; shelf-life

1. Introduction

Fresh foods are perishable foods that need to be stored under cold conditions to control microbial spoilage [1]. *Pseudomonas* genus includes bacteria that are able to grow rapidly at low temperatures and cause quality decay of different foods including milk and dairy products [2] and raw and ready to eat vegetables. Different natural antimicrobial molecules of plant, animal and microbial origin have been recently used to control spoilage bacteria in fresh foods [3]. Among animal derived compounds, the use of bovine lactoferrin (BLF) has been proposed to control microbial spoilage in different cold stored foods [4,5]. Moreover, peptides obtained by the enzymatic hydrolysis of BLF were found to delay the growth of spoiler pseudomonads in governing liquid of high moisture Mozzarella cheese [6, 7] and inhibit, *in vitro* assays, *Pseudomonas* strains in ready to eat vegetables [8]. In this work the antimicrobial efficacy of lactoferricin B (Lfcin B), obtained by enzymatic digestion (with pepsin or with cardoon (*Cynara cardunculus L).* enzymatic extract) of BLF was assayed against *Pseudomonas* spp. strains in two food model (mozzarella and lettuce) .

2. Materials and methods

2.1 Bacterial strains and culture conditions

The strains *P. fluorescens* 84095, *P. fluorescens* 84025, previously isolated from HM Mozzarella cheese [9], and *P. viridiflava* I1A, *P. fluorescens* L1A, *P. fluorescens* L1C, *P. cichorii* I3C and *P. putida* I1B [8] were used as target microorganisms for antimicrobial assays (Table 1). Each experiment was carried out using fresh growing bacterial cells obtained after incubation for 16 h at 30 °C (140 rpm) in 10 mL of Plate Count Broth (mPCB; Becton Dickinson Italia, Milan, Italy) inoculated with 50 μL of a frozen (-80 °C) culture.

2.2 Preparation of crude cardoon extract

Cardoon dried flowers (1 g) were grounded in a mortar with liquid nitrogen and suspended in 10 mL of sterile MilliQ water [10]. The suspension was vortexed for about two minutes, centrifuged at 6,000 x g for 20 min and the supernatant was filtered through 0.22 μm membrane filter (Millipore, Milan, Italy) before determining protein content by the Bradford's method [11]. The presence of aspartic proteinase, characterized by a molecular weight of ca. 30 and 15 kDa [10], was verified loading 15 μg of the cardoon extract on sodium dodecyl sulfate-polyacrylamide gel electrophoresis (SDS-PAGE, 12% T, 3% C), performed as previously reported [12]. Electrophoretic separations were carried on the Mini Protean System (Bio-Rad Laboratories; Richmond, CA,

US) at 100 V for 15 min, and 150 V for 35 min. Gels were stained with 0.125 g/mL of the Coomassie Brilliant Blue (Sigma–Aldrich, Milan, Italy), according to the method of Neuhoff et al. [13].

Table 1. Target spoilage strains.

Source	Strains	Spoilage activity
lettuce	P. fluorescens L1A	browning
	P. fluorescens L1C	browning
endive	P. cichorii I3C	browning
	P. viridiflava I1A	browning
	P.putida I1B	browning
Mozzarella cheese	P. fluorescens 84025	Blue discoloration
	P. fluorescens 84095	Blue discoloration

2.3 Production, antimicrobial assay and purification of lactoferrin pepsin-derived peptides

A 5% (w/v) bovine lactoferrin (BLF; NZMP lactoferrin 7100, Fonterra, Boulogne- Billancourt, France) solution was hydrolyzed with 3% (w/v) pepsin from porcine gastric mucosa (250 units mg^{-1} solid; Sigma-Aldrich) or crude extract cardoon (0,006% in 100 mM citrate sodium buffer, pH 5.2) as previously described [7, 10]. Bovine lactoferrin hydrolysates (LFHs) were freeze-dried and stored at -20 °C in order to be used in the subsequent experiments. To assess the hydrolysis of BLF, 10 µg of each hydrolysate was resolved on Tricine SDS-PAGE according to Quintieri et al. [7].

Target strains, cultured as described in the 2.1 section, were inoculated at 3 log cfu mL^{-1} in mPCB supplemented with 50 mg mL^{-1} of BLF or LFH. Microbial growth kinetics were evaluated by plate counts on Plate Count Agar (PCA, Biolife S.r.l., Milan, Italy) after 0, 4, 8, 24 and 30 hours of static incubation at 30 °C.

The LFH, displaying the highest antimicrobial activity, was fractionated by ion-exchange chromatography (IEX) on a Resource S column (ID 0.64 cm x 3 cm; 15 mm, GE Healthcare, Uppsala, Sweden) mounted on AktaPurifier 10 system (GE Healthcare) according to Quintieri et al. [7]. The purified peptide responsible for antimicrobial activity was freeze-dried and stored at 20 °C in order to be used in the subsequent experiments.

2.4 Application of antimicrobial peptide in fresh foods and evaluation of spoilage

Undamaged leaves from lettuce (*Lactuca sativa L.*) heads, purchased from a local market, were individually washed under tap water, disinfected for 5 min in sodium hypochlorite solution (200 ppm), rinsed in sterile distilled water at 4 °C and dried on a clean filter paper for 30 min. Then, small pieces of leaves (midrib length of 5-6 cm) were transferred in Petri dishes and incised (50 incisions/piece) along the central rib by using a sterile scalpel. Freeze-dried powder of the purified peptide was dissolved at 3 mg mL^{-1} in the sterile saline solution and in bacterial suspension (3 log cfu mL^{-1}) of the selected strain, respectively. Treatments were performed adding to each incision 10 µL of sterile saline solution, peptide solution, bacterial suspension or bacterial suspension amended with the antimicrobial peptide. The infection percentage was calculated until 5 days at 4 °C, recording the number of incisions that displayed evident pigmentations (yellowing, browning or reddish pigmentation) in comparison with the total incisions within each treatment.

Fresh Mozzarella cheeses, purchased from a local dairy farm and produced by means microbial acidification, were sliced under sterile condition and spotted (4 spots/slice) with 10 µL of a filter-sterilized (0.22-mm-pore size, Millipore) peptide solution (3 mg mL^{-1}) or 0.95% NaCl solution containing 3 log cfu mL^{-1} of the target strain. Mozzarella cheese slices with the same volume of un-inoculated sterile saline solution were included in the assay as a control. All samples were prepared in triplicate and incubated at 4 °C for a week.

3. Results and discussion

3.1 Electrophoretic characterization of thistle cardoon extract

Aqueous extracts of *Cynara cardunculus* have been used as milk coagulant in the manufacture of cheeses [14] thanks to its milk-clotting activity accounted by two aspartic proteinases, named cardosins A and B. Since the activity and specificity of these proteins resemble those found for chymosin and pepsin [10], *Cynara cardunculus* extracts hydrolysed BLF releasing antimicrobial peptides as already demonstrated for pepsin [15].

The total protein content of the crude extract from the thistle flower was 1.8 mg mL^{-1}. The resultant SDS-PAGE electrophoretic pattern showed polypeptides with molecular weights of ca. 30 kDa and 15 kDa, corresponding putatively to the subunits of cardosin A (30 and 15 kDa) and B (31 and 14 kDa), respectively (Figure 1). This result was in accordance to that reported by other Authors [10].

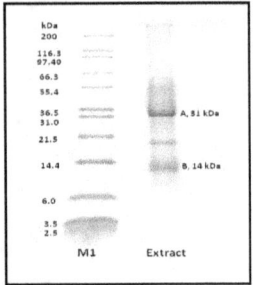

Fig. 1. SDS-PAGE (T: 12 %; C: 3%) profile of an aqueous flower extract from cardoon (*Cynara cardunculus*); M1, Invitrogen M12^{TM} (200.00 -2,5 kDa).

3.2 Production of lactoferrin pepsin-derived peptides and antimicrobial assay

The protein pattern of commercial BLF used in this work was mainly made up of lactoferrin (ca. 80 kDa) together with 4 unknown proteins or peptides (Fig. 2). After 4 hours of incubation, each enzyme was able to digest BLF showing the complete disappearance of the related band. Hydrolysates obtained by means of digestion with cardoon crude extract also showed bands with a molecular weight higher than 14.0 kDa (Fig. 2B). In contrast, the hydrolysis of lactoferrin with pepsin released three major peptides with apparent molecular weight of ca 3, 5 and 14 kDa (Fig. 2A), as previously reported by Quintieri et al. [7].

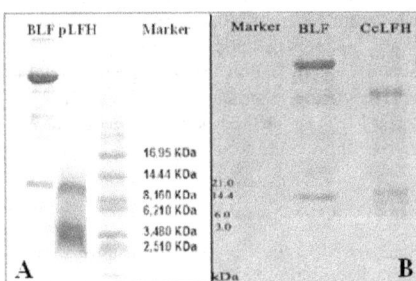

Fig. 2. Tricine SDS-PAGE profiles of bovine lactoferrin hydrolysates after 60 min of digestion. **A,** BLF, bovine lactoferrin; pLFH, pepsin digested bovine lactoferrin; Marker, Peptide Molecular Weight Marker (Biorad). **B,** Marker, Peptide Molecular Weight Marker (Sigma); BLF, bovine lactoferrin; CcLFH, bovine lactoferrin hydrolysate obtained by digestion with crude cardoon extract.

In order to reduce the production of off-colours by several target spoiler strains in two fresh food models (vegetable and cheese), lactoferrin hydrolysates from pepsin and thistle cardoon extracts (pLFH and CcLFH, respectively) were evaluated for their *in vitro* antimicrobial activity. After 30 hours of incubation in PCB, all target strains increased in viable cell loads, reaching average values of 8.3 log cfu mL^{-1} (Table 2), whereas the cultures of the same strains containing pLFH were completely inhibited until 30 hours of incubation (Table 2). Moreover, the inhibited cultures, inoculated (1%; v/v) in fresh PCB without any antimicrobials displayed no residual growth until 48 hours of incubation. Similar results were found when CcLFH was used. Recently, the efficacy of pLFH was proved *in vitro* against *Pseudomonas* spp. isolated from cold stored Mozzarella cheese [6] and able to grow at 4°C and promote caseinolytic activities in Mozzarella cheese [2].

Starting from 5 hours and up to the end of incubation, pLFH was proved to inhibit significantly the growth of 3 out of the 5 selected strains; in fact, *P. chicorii* I3C and *P. viridiflava* I1A counts decreased of ca. 4.5 log cycle, on average, in comparison with the untreated cultures, whilst no cells were enumerated for *P. putida* I1B (Table 2). This latter strain was also sensitive to CcLFH, registering a reduction by 6.7 log cfu mL^{-1} at 30 hours of incubation. In contrast, the remaining strains *P. fluorescens* L1A and L1C were resistant to both peptide mixtures (data not shown). The native bovine lactoferrin showed a weak or no antimicrobial activity towards all the selected strains, confirming the results previously reported by several Authors on the higher antibacterial activity of its peptides [15,16,7].

Since pLFH showed the highest inhibition activity against *Pseudomonas* spp spoilers compared to that of CcLFH further experiments were performed only with pLFH.

Table 2. Microbial counts (log cfu mL^{-1}) of indicator strains grown in PCB at 30 °C up to 30 hours in presence of 50 mg mL^{-1} of pepsin-digested lactoferrin hydrolysate (pLFH) or bovine lactoferrin hydrolysate obtained by digestion with crude cardoon extract (CcLFH) or without pLFH and CcLFH (none).

Strains	Hydrolysate	Hours of incubation			
		0	**5**	**24**	**30**
P. cichorii I3C	none	4.28 ± 0.16	4.06 ± 0.16	8.16 ± 0.13	8.63 ± 0.11
	CcLFH	4.28 ± 0.10	4.72 ± 0.60	8.63 ± 0.05	8.28 ± 0.16
	pLFH	3.91 ± 0.19	3.31 ± 0.12	4.20 ± 0.26	4.79 ± 0.17
P. viridiflava I1A	none	3.90 ± 0.26	4.43 ± 0.12	10.63 ± 0.25	10.56 ± 0.11
	CcLFH	3.90 ± 0.26	4.67 ± 0.36	8.95 ± 0.16	9.02 ± 0.09
	pLFH	3.90 ± 0.26	3.55 ± 0.15	5.80 ± 0.16	4.96 ± 0.41
P. putida I1B	none	4.37 ± 0.16	4.46 ± 0.18	9.72 ± 0.16	9.04 ± 0.16
	CcLFH	4.37 ± 0.16	1.00 ± 0.36	1.00 ± 0.44	2.33 ± 0.56
	pLFH	4.37 ± 0.16	-	-	-
P. fluorescens 84095	none	3.22 ± 0.56	3.88 ± 0.36	8.71 ± 0.16	8.71 ± 0.19
	CcLFH	3.22 ± 0.56	-	-	-
	pLFH	3.22 ± 0.56	-	-	-
P. fluorescens 84025	none	3.94 ± 0.23	3.71 ± 0.62	9.47 ± 0.16	8.03 ± 0.12
	CcLFH	3.94 ± 0.23	-	-	-
	pLFH	3.94 ± 0.23	-	-	-

3.3 Application of antimicrobial peptides in fresh vegetable and cheese models and evaluation of spoilage

Recently, in a previous work we assessed [7] that the antibacterial activity of pLFH was mainly correlated to the release of the peptide Lactoferricin B (LfcinB $_{17\text{-}42}$). This peptide showed a wide inhibitory activity against several *Pseudomonas* strains counteracting their proteolytic ability. Thus, in this work the control of pigmentation caused by spoiler *Pseudomonas* spp strains on selected lettuce and Mozzarella samples was evaluated by using LfcinB purified from pLFH hydrolysate (Figs. 3-4, Table 3).

Starting from the day 3 of cold storage, cheese slices inoculated with *P. fluorescens* 84095 showed a greenish/bluish color discoloration, whereas the inoculated samples treated with LfcinB appeared similar in color to the uninoculated Mozzarella cheese.

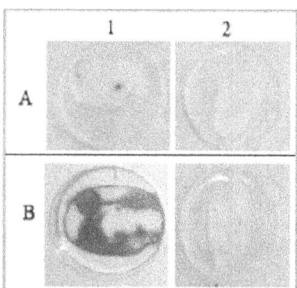

Fig. 3. Effect of treatment of lactoferricinB (LfcinB) and storage time on blue discoloration of Mozzarella cheese slices incubated at 4 °C for 7 days. **A**, Mozzarella cheese slices at day 3, inoculated with *Pseudomonas fluorescens* 84095 (1) or inoculated and supplemented with LfcinB (2); **B**, Mozzarella cheese slice at day 7 inoculated with *Pseudomonas. fluorescens* 84095 (1) or inoculated and supplemented with LfcinB (2).

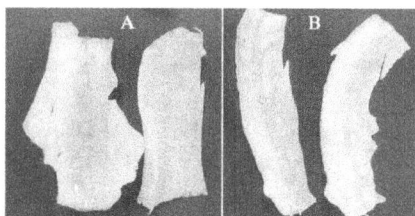

Fig. 4. Effect of treatment of lactoferricinB (LfcinB) on browning of incised leaves at 4 °C for 120 hours. **A**, incised leaves inoculated with *P. putida* I1B; **B**, incised leaves inoculated with *P. putida* I1B and treated with Lfcin B.

Furthermore, a significant reduction of wound browning was observed in ready-to-eat lettuce leaves inoculated with different *Pseudomonas* spoilage strains and treated with LfcinB (Table 3). On the other hand, high percentages of browned wounds were found on LfcinB-free inoculated lettuce leaves. Nevertheless, the treatment with LfcinB did not counteract the residual pigmentation on LfcinB-free and uninoculted wounds ought to enzymatic pathways associated with the wounding of vegetable tissue.

This work demonstrates the possibility to apply food-grade antimicrobial peptides derived from BLF to hinder spoilage of the ready-to-eat vegetables and fresh cheeses and extend their shelf-life.

Table 3. Effect of lactoferricinB (LfcinB) on the development of browning caused by *Pseudomonas* strains.

Strains	Treatments	Percentage of browned wounds (%)			
		Cold storage (hours)			
		72	96	120	144
P. putida I1B	Inoculated	54	90	96	100
	Inoculated+ LfcinB	26	36	44	64
P. viridiflava I1A	Inoculated	65	68	70	86
	Inoculated +LfcinB	30	44	48	58
P. chicorii I3C	Inoculated	48	85	100	100
	Inoculated+ LfcinB	30	44	48	53
None	Uninoculated	30	34	38	53
	Uninoculated +LfcinB	27	32	37	52

Acknowledgments: This work was carried out under research activities of the Project: "High-Convenience Fruits and vegetables: New Technologies for Quality and New Products" (OFRALSER). PON01_01435. financed by the Italian Ministry of Education University and Research.

References

[1] Sperber WH. Introduction to the Microbiological Spoilage of Foods and Beverages. In: Sperber WH, Doyle MP, editors. Compendium of the Microbiological Spoilage of Foods and Beverages. New York: Springer, 2010. p. 1-40.

[2] Baruzzi F, Lagonigro R, Quintieri L, Morea M, Caputo L. Occurrence of non-lactic acid bacteria populations involved in protein hydrolysis of cold stored high moisture mozzarella cheese. Food Microbiology. 2012; 30: 37-44.

[3] Juneja VK, Dwivedi HP, Yan X. Novel Natural Food Antimicrobials. Annual review of food science and technology. 2012; 3: 381-403.

[4] Umuhumuza LC, Wei-min N, Sun X. Effect of Bovine lactoferrin and casein peptide powder on microbial growth and glucose utilization by microorganisms in pork meat during storage at 4 C. Pakistan Journal of Nutrition. 2011; 10(3): 208-213.

[5] Chiu CH, Kuo CC. antioxidative and antimicrobial properties of lactoferrin in hot-boned ground pork during storage. Journal of food processing and preservation. 2007; 31(2), 157-166.

[6] Quintieri L, Caputo L, Morea M, Baruzzi F. Control of Mozzarella spoilage bacteria by using bovine lactoferrin pepsin-digested hydrolysate. In: A. Mendèz-Vilas editor: Worldwide Research Efforts in the Fighting against Microbial Pathogens: From Basic Research to Technological Developments. Boca Raton: BrownWalker Press. 2013. p. 118-122.

[7] Quintieri L, Caputo L, Monaci L, Deserio D, Morea M, Baruzzi F. Antimicrobial efficacy of pepsin-digested bovine lactoferrin on spoilage bacteria contaminating traditional mozzarella cheese. Food Microbiology. 2012. 31(1): 64–71.

[8] Carito A, Quintieri L., Pinto L., Calabrese N., Baruzzi F. Caputo L. Influenza della lattoferrina e dei suoi peptidi sulla crescita *in vitro* di microrganismi alterativi di ortaggi di IV gamma. Proceedings of the PostHarvest Conference, 2014 May 22-24, Barletta, Italy.

[9] Caputo L, Quintieri L, Bianchi DM, Decastelli L, Monaci L, Visconti A, Baruzzi F. Pepsin-digested bovine lactoferrin prevents Mozzarella cheese blue discoloration caused by *Pseudomonas fluorescens*. Food Microbiology. 2015; 46:15-24.

[10] Silva SV, Allmere T, Malcata FX, Andren A. Comparative studies on the gelling properties of cardosins extracted from *Cynara cardunculus* and chymosin on cow's skim milk. International Dairy Journal. 2003; 13, (7): 559-564.

[11] Bradford MM. A rapid and sensitive method for the quantitation of microgram quantities of protein utilizing the principle of protein-dye binding. Analitycal Biochemistry. 1976. 72: 248-254.

[12] Laemmli UK. Cleavage of structural proteins during the assembly of the head of bacteriophage T4. Nature. 1970. 227:680–685.

[13] Neuhoff V, Arold N, Taube D, Ehrhardt W. Improved staining of proteins in polyacrylamide gels including isoelectric focusing gels with clear background at nanogram sensitivity using Coomassie brilliant blue G-250 and R-250. Electrophoresis. 1988. 9: 255-262.

[14] Vieira de Sâ F, Barbosa M. Cheese-making with a vegetable rennet from cardoon (*Cynara cardunculus*). Journal of Dairy Research. 1972. 39: 335-34.

[15] Tomita M, Bellamy W, Takase M, Yamauchi K, Wakabayashi H, Kawase K. Potent antibacterial peptides generated by pepsin digestion of bovine lactoferrin. Journal of Dairy Science. 1991. 74: 4137-4142.

[16] Wakabayashi H, Takase M, Tomita M. Lactoferricin derived from milk protein lactoferrin. Current Pharmaceutical Design. 2003. 9: 1277-1287.

Bactericidal effect of encapsulated caprylic acid on *Listeria monocytogenes*

M. Ruiz-Rico[1,2], **E. Pérez-Esteve**[1,2], **A. Jiménez-Belenguer**[3], **M.A. Ferrús**[3], **R. Martínez-Mañez**[2,4] **and J.M. Barat**[1]

[1] Grupo de Investigación e Innovación Alimentaria, Departamento de Tecnología de Alimentos, Universitat Politècnica de València, Camino de Vera s/n., 46022 Valencia, Spain

[2] Centro de Reconocimiento Molecular y Desarrollo Tecnológico, Unidad Mixta Universitat Politècnica de València-Universitat de València, Camino de Vera s/n, 46022, Valencia, Spain

[3] Departamento de Biotecnología, Universitat Politècnica de València, Camino de Vera s/n., 46022 Valencia, Spain

[4] CIBER de Bioingeniería, Biomateriales y Nanomedicina (CIBER-BBN)

*Corresponding author: e-mail: anjibe@upvnet.upv.es, Phone: +34 963877423

Mesoporous silica particles (MSPs) are considered as promising vehicles for the controlled delivery of bioactive molecules from food systems. Natural antimicrobial compounds such as free fatty acid could be encapsulated in this type of support in order to protect their antimicrobial activity against degradation processes. This encapsulation system has been recently reported, showing promising results as antimicrobial delivery supports. In the present study we proposed two different MSP (MCM-41 nano- and microparticle) for the encapsulation of caprylic acid (CA) to establish the minimum bactericidal concentration of the entrapped fatty acid against *Listeria monocytogenes*. Results show that entrapped fatty acid produced a reduction of microbial population in both encapsulation systems. The most effective encapsulated condition was found for nanoparticles being the minimum bactericidal concentration of CA encapsulated in this support in the range of 18.5 – 20 mM. Otherwise, CA encapsulated in microparticles produced a reduction of microbial growth, but this encapsulation system was not bactericidal for any of the studied concentrations. Furthermore, bacterial viability assays were done with unloaded MSPs to test the possible toxicity of the MSPs. Results show that unloaded MSPs did not affect the population counts of the microorganism. These results suggest that MSP supports are not toxic to bacteria and the effect of loaded solids is due to the fatty acid delivery. Therefore, the encapsulation process allows maintaining the antimicrobial activity of CA and could provide a new system to guarantee food safety in food industry using a more stable form of administration.

Keywords caprylic acid; minimum bactericidal concentration; encapsulation; mesoporous silica particles

1. Introduction

Synthetic preservatives agents are chemical compounds able to kill or slow down the growth of bacteria without coursing toxicity to animals or humans, and they are used in food, packaging or contact surfaces in order to extend the shelf life of food products and control the growth of foodborne bacteria. Despite the effectiveness of these compounds, consumers are demanding for natural antimicrobial compounds or new administration forms to reduce health problems [1].

Among different natural antimicrobial compounds, free fatty acids (FFAs) are known to be potent antimicrobial agents against enveloped viruses, bacteria and fungi [2]. One saturated FFA that it is used as natural antimicrobial agent is CA, also known by the systematic name octanoic acid, and it has a chain length of eight-carbon. CA is found in breast milk of various mammals and is a minor component of coconut oil and palm kernel oil [2, 3]. It is a food-grade compound classified as generally recognized as safe (GRAS) by the U.S. Food and Drug Administration, and considered safe when used as a flavour by the Joint FAO/WHO Expert Committee on Food Additives [4]. The effectiveness of free CA against foodborne pathogens *Escherichia coli*, *Listeria monocytogenes*, *Cronobacter* spp., and S*almonella* spp. in food has been reported [2, 3, 5-7]. Moreover, in some in vivo studies it has been proven that CA significantly decreases high mortality of growing rabbits, without affecting the rate of growth, feed intake and carcass yield [8].

The application or supplementation with CA could produce some disadvantages derived for its sensorial properties such as the unpleasant rancid-like smell and taste [9]. These problems may be prevented by the use of an encapsulation support, which, moreover, allows the FFA to maintain its properties, increase the stability against external agents and reduce the effective dose.

Among different encapsulation techniques, mesoporous silica particles (MSPs) exhibit unique features as supports for controlled release, such as high stability, biocompatibility, no apparent toxicity, large load capacity and the possibility to include gate-like scaffoldings on the external surface [10-12]. The encapsulation of natural antimicrobial compounds in this type of particles has been recently reported [13], showing promising results as a delivery support.

The purpose of this study was to evaluate the antimicrobial activity of caprylic acid encapsulated in mesoporous silica supports against *Listeria monocytogenes*.

2. Materials and methods

2.1 Mesoporous silica particles synthesis

Nanoparticulated MCM-41 particles (N) were synthesized using the molar ratio 1 TEOS:0.1 CTABr:0.27 NaOH:1000 H2O. NaOH was added to the CTABr solution, followed by adjusting the solution temperature to 95 °C. TEOS was then added dropwise to the surfactant solution. The mixture was allowed to stir for 3 h yielding a white precipitate. Synthesis of microparticulated MCM-41 particles (M) was carried out using CTABr as the structure-directing agent and a molar ratio fixed to 7 TEAH3: 2 TEOS:0.52 CTABr:0.5 NaOH:180 H2O. CTABr was added to a solution of TEAH3 and NaOH containing TEOS at 118 °C. After dissolving CTABr in the solution, water was slowly added with vigorous stirring at 70 °C to form a white suspension which was aged at room temperature overnight.

After the synthesis, the different solids were recovered, washed with deionised water, and air-dried at room temperature. The as-synthesized solids were calcined at 550 °C using an oxidant atmosphere for 5 h in order to remove the template phase.

For the loading of the particles, 200 mg of free-template solid were suspended in a solution of 5.7 g of CA in 19 mL of n-hexane in a round-bottomed flask and the mixture was stirred for 24h at room temperature. The loaded solids were isolated by vacuum filtration and dried at room temperature for 24h.

2.2 Materials characterization

MSPs were characterized through standard techniques: transmission electron microscopy (TEM), particle size distribution and zeta potential. For TEM analysis, MSPs were dispersed in dichloromethane and the suspension was deposited onto copper grids coated with carbon film (Aname SL, Madrid, Spain). Imaging of the MSPs samples was performed using a JEOL JEM-1010 (JEOL Europe SAS, France) operating at an acceleration voltage of 80 kV. Microscopy studies were completed by determining size distribution of solids dispersed in Tryptic Soy Broth (TSB) with a Malvern Mastersizer 2000 (Malvern, UK). Data analysis was based on the Mie theory using refractive indices of 1.33 and 1.45 for the dispersant and MSP respectively. Measurements were performed in triplicate. Zeta potential of different solids dispersed in TSB was measured in a Zetasizer Nano ZS equipment (Malvern Instruments, UK). The zeta potential was calculated from the particle mobility values by applying the Smoluchowski model. The average of five recordings is reported as zeta potential.

2.3 Microbiological analysis

2.3.1 Culture media and bacterial culture

Plate Count agar and TSB were used to grow bacteria and prepare inoculums. Palcam agar base supplemented with polymyxin B, acriflavine and ceftazidime, was used to grow *L. monocytogenes*. All the mediums were provided by Scharlau (Barcelona, Spain). *L. monocytogenes* (CECT 936) strain was obtained from the Colección Española de Cultivos Tipo (CECT) of Valencia (Spain). The bacterial stock was stored at 4 °C in solid medium before use. Bacteria cells were grown aerobically in TSB at 37 °C for 24 h.

2.3.2 Bacterial viability assays

The bacterial viability of the microorganism was tested with concentrations of unloaded nano- and microparticles suspended in the nutrient broth. These concentrations (1, 2.5, 5 and 10 mg mL^{-1}) are in the range of the amount of loaded solids needed to perform the antimicrobial susceptibility assays.

To achieve the final particle concentrations 0.04, 0.1, 0.2 and 0.4 mL of 250 mg mL^{-1} particle stock suspension were added to 10 mL of TSB in test tubes. They were inoculated with 10 μL of inoculum and incubated under orbital stirring (150 rpm) at 37 °C for 24 h. All treatments were set in triplicate.

To determine bacterial viability, CFUs were counted by plating serial dilutions of bacterial suspensions on selective agar, followed by incubation at 37 °C for 48 h. The results were expressed as log CFU mL^{-1}. The control positive values were used to quantify the percentage of survival of the microorganisms for the different studied concentrations.

2.3.3 Antimicrobial susceptibility assays of loaded particles

The antimicrobial activity of CA entrapped in MSPs was carried out using the concentrations of CA 15, 18.5, 20, 22.5 and 25 mM. The required amount of solids was calculated from the actual released concentration of CA and were added to 10 mL of TSB. The test tubes were inoculated with 10 μL of inoculum and incubated. All treatments were set in triplicate.

To quantify bactericidal activity, serial dilutions of the incubation mixtures were plated at 24 and 48 h of incubation, followed by incubation at 37 °C for 48 h. The results were expressed as log CFU mL^{-1}. The control positive values were used to quantify the percentage of growth reduction for the different tested conditions.

In all the microbiological assays, a positive control (test tube containing inoculum and nutrient broth without CA or MSPs) and a negative control (test tube containing CA or MSPs and nutrient broth, without bacteria) were included. The positive control indicated the bacterial growth profile in absence of antimicrobial compound and the negative control confirmed the absence of contamination of MSPs.

2.4 Data analysis

Statistical treatment of the data was performed using Statgraphics Centurion XVI (Statpoint Technologies, Inc., Warrenton, VA, USA). An analysis of variance (One-way ANOVA) was done in order to evaluate the effect of CA concentration and exposure time on the reduction of the growth of the microorganism. The LSD procedure (least significant difference) was used to test for differences between averages at the 5% significance level.

3. Results and discussion

3.1 Characterization of mesoporous silica particles

The different solids prepared were characterized according to standard techniques. The MCM-41 structure after calcination was confirmed by TEM (Figure 1).

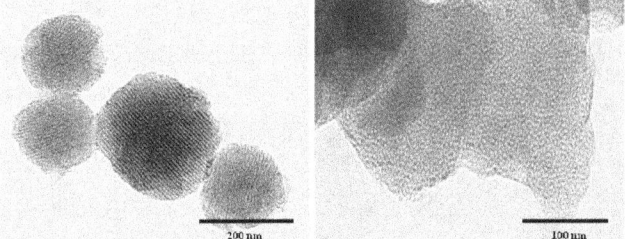

Fig. 1 TEM images of the inorganic MCM-41 nanoparticle calcined matrix N1 (left) and calcined microparticle M1 (right).

The structure of the MCM-41 matrix as alternate black and white stripes can be observed in Figure 1. Besides porosity, the morphologic analysis with TEM allowed the determination of size of nano- and microparticles.

Due to particles that were employed in an aqueous media, particle size obtained by TEM were compared with particle size obtained by Light Diffraction (Table 1).

When single particle sizes and size distribution are compared it can be seen that particle size increased dramatically in the case of nanoparticles, passing from the nanoscale to the microscale, in the same way that previous studies [12]. To explain changes of particle size, zeta potential determinations were performed. Both types of particles exhibited zeta potential values next to -15 mV (range of the instability) and this could be the reason because particles increased its size as a consequence of particle aggregation.

Table 1 Particle size of different MCM-41 particles determined by TEM (dry) or Light Diffraction (dispersed in TSB), and zeta potential. (Means and standard deviations).

Particle	Particle size (μm)		Zeta potential (mV)
	TEM	d (0,5)	
Nanoparticle	0.088 ± 0.013	1.580 ± 0.001	-17.05 ± 2.30
Loaded nanoparticle	0.114 ± 0.0183	1.634 ± 0.001	-14.68 ± 0.92
Microparticle	1.129 ± 0.493	1.484 ± 0.001	-14.67 ± 1.68
Loaded microparticle	1.037 ± 0.516	1.619 ± 0.013	-14.32 ± 0.69

3.2 Effect of mesoporous silica matrix on bacterial viability

The bacterial viability of the microorganism was tested with concentrations of nano- and microparticles at concentrations 1-10 mg mL^{-1} suspended in the nutrient broth. As it can be seen in Figure 2 a survival close to 100% of the microorganism was obtained for all the tested concentrations, which indicates that the treatment with unloaded particles did not affect the microbial growth of any bacteria.

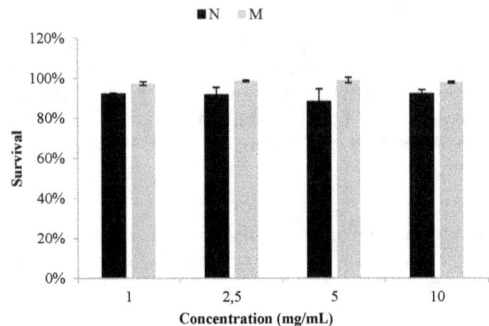

Fig. 2 Bacterial viability of *L. monocytogenes* against MCM-41 microparticle (M) and nanoparticle (N).

These results are in accordance with a previous study, in which the effect of silica particles with size between 15 and 500 nm on bacterial viability was investigated and it was established that the particles displayed no inhibitory properties independently from their particle size [14].

3.3 Inhibitory activity of entrapped CA in nano- and microparticle MCM-41

The inhibitory effect of the loaded nanoparticles and microparticles against *L. monocytogenes* at 24 and 48 h of incubation is shown in Figure 3.

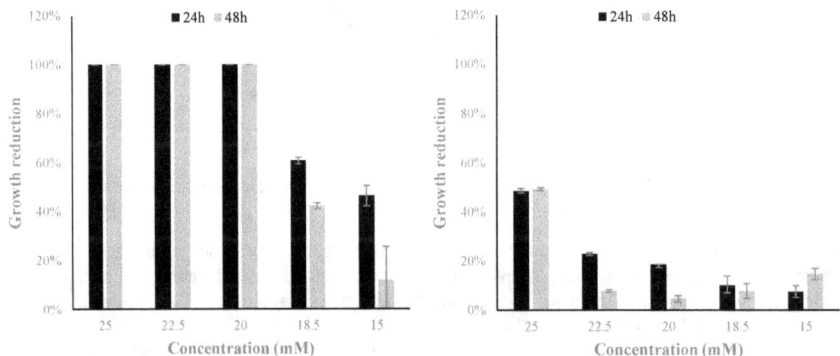

Fig. 3 Reduction of microbial growth (%) of *L. monocytogenes* under loaded nanoparticles (left) and loaded microparticles (right).

The bactericidal effect of CA entrapped in nanoparticles was between 18.5 – 20 mM at different times of treatment. Otherwise, lower concentrations reduced more the microbial growth at time of incubation 24 h than 48 h, showing statistically differences for this factor in the concentration 18.5 mM. These results are in agreement with other studies that established a minimum inhibitory concentration between 5-25 mM for *L. monocytogenes* after incubation in nutrient broth or milk [5, 15]. In the case of loaded microparticles, none of the studied concentrations was bactericidal for the microorganism. The maximum reduction of the bacterial growth was achieved by the concentration of 25 mM reaching a value of 49.219 ± 0.006 % on average for both times of incubation.

4. Conclusions

Natural antimicrobial compounds and new administration forms could be a good alternative to traditional antimicrobial agents. The encapsulation allows reducing the effective doses of the antimicrobial compounds due to the control release of the target substance, and it protects the stability and functionality of the antimicrobial compound. This study had showed the effectiveness of the CA as antimicrobial compound and the suitability of MSPs to entrap the CA and remain its antimicrobial activity. Nanoparticles MCM-41 were the most appropriate support for CA encapsulation showing a MBC in the range of 18.5 – 20 mM. Nanoencapsulation of CA may be an attractive option for food industry to guarantee food safety, and these results represent a new line of research for the use of controlled release systems based on nanoparticles. Further studies on encapsulation and microbial susceptibility of several strains will be necessary to validate this procedure.

Acknowledgements Authors gratefully acknowledge the financial support from the Ministerio de Economía y Competitividad (Projects AGL2012-39597-C02-01, AGL2012-39597-C02-02). M.R.R. and E.P. are grateful to the Ministerio de Ciencia e Innovación for their grants (AP2010-4369 and AP2008-0620).

References

[1] Weiss, J, Gaysinsky, S, Davidson, M, McClements, J. Nanostructured encapsulation systems: food antimicrobials. In IUFoST world congress book: Global issues in food science and technology. 2009; 425-479.
[2] Chang, S, Redondo-Solano, M, Thippareddi, H. Inactivation of *Escherichia coli* O157:H7 and *Salmonella* spp. on alfalfa seeds by caprylic acid and monocaprylin. International Journal of Food Microbiology. 2010; 144:141–146.
[3] Jang, HI, Rhee, MS. Inhibitory effect of caprylic acid and mild heat on Cronobacter spp. (Enterobacter sakazakii) in reconstituted infant formula and determination of injury by flow cytometry. International Journal of Food Microbiology. 2009; 133:113–120.
[4] JECFA (Joint FAO & WHO Expert Committee on Food Additives). Evaluation of certain food additives and contaminants. World Health Organization technical report series. 1999; 896:1.
[5] Nair, MKM., Vasudevan, P, Hoagland, T, Venkitanarayanan, K. Inactivation of Escherichia coli O157:H7 and *Listeria monocytogenes* in milk by caprylic acid and monocaprylin. Food Microbiology. 2004 ; 21 :611–616.
[6] Vasudevan, P, Marek, P, Nair, MKM, Annamalai, T, Darre, M, Khan, M, Venkitanarayanan, K. In Vitro Inactivation of *Salmonella enteritidis* in Autoclaved Chicken Cecal Contents by Caprylic acid. Poultry Science Association. 2005; 14(1):122-125.
[7] Choi, MJ, Kim, SA, Lee, NY, Rhee, MS. New decontamination method based on caprylic acid in combination with citric acid or vanillin for eliminating *Cronobacter sakazakii* and *Salmonella enterica* serovar *Typhimurium* in reconstituted infant formula. International Journal of Food Microbiology. 2013; 166:499–507.
[8] Skrivanová, E, Molatová, Z, Marounek, M. Effects of caprylic acid and triacylglcerols of both caprylic and capric acid in rabbits experimentally infected with enteropathogenic *Escherichia coli* O103. Veterinary Microbiology. 2008; 126: 372–376.
[9] Hulankova, R, Borilova, G, Steinhauserova, I. Combined antimicrobial effect of oregano essential oil and caprylic acid in minced beef. Meat Science. 2013; 95:190–194.
[10] Slowing, II, Vivero-Escoto, JL, Wu, CW, Lin, VSY. Mesoporous silica nanoparticles as controlled release drug delivery and gene transfection carriers. Advanced Drug Delivery Reviews. 2008; 60:1278-1288.
[11] Al Shamsi, M, Al Samri, MT, Al-Salam, S, Conca, W, Shaban, S, Benedict, S, Tariq, S, Biradar, AV, Penefsky, HS, Asefa, T, Souid, A-K. Biocompatibility of calcined mesoporous silica particles with cellular bioenergetics in murine tissues. Chem. Res. Toxicol. 2010; 23:1796–1805.
[12] Pérez-Esteve, E, Oliver, L, García, L, Nieuwland, M, de Jongh, HH, Martínez-Máñez, R, Barat, JM. Incorporation of Mesoporous Silica Particles in Gelatine Gels: Effect of Particle Type and Surface Modification on Physical Properties. Langmuir. 2014; 30:6970–6979.
[13] Park, SY, Pendleton, P. Controlled release of allyl isothiocyanate for bacteria growth management. Food control. 2012; 23:478-484.
[14] Wehling, J, Volkmann, E, Grieb, T, Rosenauer, A, Maas, M, Treccani, L, Rezwan, K. A critical study: Assessment of the effect of silica particles from 15 to 500 nm on bacterial viability. Environmental Pollution. 2013; 176:292-299.
[15] Nobmann, P, Smith, A, Dunne, J, Henehan, G, Bourke, P. The antimicrobial efficacy and structure activity relationship of novel carbohydrate fatty acid derivatives against *Listeria* spp. and food spoilage microorganisms. International Journal of Food Microbiology. 2009; 128, 440–445.

Cinnamic acid in the control of planktonic and sessile cells of *Escherichia coli* and *Staphylococcus aureus*

J. Malheiro[1], **I. Gomes**[1], **A. Borges**[1], **A. Abreu**[1], **J. Loureiro**[1], **F. Mergulhão**[1] and **M. Simões**[1,*]

[1] LEPABE, Department of Chemical Engineering, Faculty of Engineering, University of Porto, Rua Dr. Roberto Frias, s/n, 4200-465 Porto, Portugal
*Corresponding author: e-mail: mvs@fe.up.pt, Phone: +351 22 508 1654

Phytochemicals (plant secondary metabolites) are a source of new antimicrobial products. In this this study, cinnamic acid and ferulic acid, two phytochemical products were tested for their antimicrobial, antiadhesive and antibiofilm properties against *Escherichia coli* and *Staphylococcus aureus*. Their effects were compared with two in-use biocides, sodium hypochlorite and hydrogen peroxide. The most efficient chemical was sodium hypochlorite with almost 100% bacterial inactivation in both adhered and biofilm forms. Cinnamic acid had antimicrobial activity and an antiadhesive effect comparable to those of sodium hypochlorite. In addition, a mild bacteria inactivation was accomplished in biofilms. This study reinforces that phytochemicals are an auspicious alternative to commonly used disinfectants for general disinfection practices.

Keywords Antimicrobial resistance; Cinnamic acid; Disinfection; Ferulic acid; Phytochemicals

1. Introduction

Surfaces and inanimate objects are a great concern for pathogens transmission in hospital settings [1].

Health care-acquired infections caused by methicillin-resistant and susceptible *Staphylococcus aureus* (MRSA and MSSA), resistant gram-negative bacteria (MRGN) and *Clostridium difficile* are related with high morbidity and mortality [1, 2]. In hospitals, surfaces which often experience hand contact are frequently contaminated with nosocomial pathogens that may serve as vectors for cross-transmission [3]. The high hazard is mainly due to the fact that nosocomial pathogens can persevere on surfaces and objects for weeks or even months or on fingers for up to several hours [1]. Regarding hospital settings, the surface contamination rates can reach 58% in the case of *C. difficile*, 1 to 18% (nonburn wards) or 64% (burn units) for MRSA and 37% for vancomycin-resistant enterococci (VRE) [4]. These facts together with the known outbreaks that result from improperly decontaminated patient-care items highlight the necessity for and appropriate disinfection procedure [5]. The decontamination of hospital surfaces is essentially accomplished by disinfection that typically uses a chemical agent that is normally able to eliminate the majority of the microorganism, excluding bacterial spores [4].

However, the increase in the use of disinfectants without proper application and usage has led to a reduction in bacterial susceptibility by several mechanisms, such as phenotypic adaptation, genetic mutations or genetic acquisition [6, 7]. Furthermore, due to evidence of a strong relationship between disinfectant and antibiotic resistance, there is a real risk in increasing clinically relevant organism resistance [8]. This problem imposes an enormous health and economic burden. For example, regarding the problem of antibiotic-resistant infections, Europe Union presented the burden imposed yearly: 1.5 billion € and 25000 associated deaths [9].

The growing number of resistant bacteria to in-use disinfectants imposes a higher necessity for the research of new sources of chemicals with antimicrobial activity. In the last decades phytochemicals, specially phenolic compounds have been extensive explored for their antimicrobial properties. This group has chemicals with a wide range of characteristics, such as antibacterial, antiallergic, anti-inflamatory, antirhombi, hepatoprotective, antiviral, anticarcinogenic and vasodilatory action [10, 11].

In this study, two phytochemicals, cinnamic acid and ferulic acid, were tested for their antimicrobial properties, against planktonic and sessile cells, in comparison with two disinfectants widely used in health settings, sodium hypochlorite and hydrogen peroxide [12, 13].

2. Materials and Methods

2.1 Antimicrobial agents and microorganisms

Cinnamic acid and hydrogen peroxide were purchased from Merck (VWR, Portugal). Ferulic acid was purchased from Sigma (Portugal). Sodium hypochlorite was purchased from Acros Organics (Portugal). All dilutions were performed using dimethyl sulfoxide (DMSO) for phytochemicals or sterile distilled water for disinfectants.

The bacteria used in this study were *Escherichia coli* CECT 434 and *Staphylococcus aureus* CECT 976.

2.2 Antibacterial susceptibility testing – minimum inhibitory and bactericidal concentrations

The minimum inhibitory concentration (MIC) of each agent was determined by the microdilution method according to the Clinical and Laboratory Standards Institute (CLSI) guidelines [14]. The MIC was determined as the lowest concentration that inhibited microbial growth [15]. A volume of 10 μl/well was plated in Plate Count Agar (PCA, Merck, Germany) and incubated overnight at $30 \pm 3°C$, after a neutralization step to quench the chemicals antimicrobial activity, by dilution to sub-inhibitory concentrations [16]. The MBC was considered the lowest concentration of the antimicrobial chemical were no growth was detected on the solid medium [15].

2.3 Adhesion and biofilm formation in 96-well polystyrene plates

A bacterial suspension was dispersed to a 96-well polystyrene plate (200 μl/well) and adhesion was allowed to occur for 2 hours (adhesion) or 24 hours (biofilm formation) at 30 °C and 150 rpm, according to Simões et al (2007) [17]. Then the planktonic cells were discarded and the cells are washed with 8.5 g/L NaCl prior the addition of the chemical (10% v/v of chemical dissolved in NaCl solution (8.5 g/L)). The different chemicals were tested at 10 mM. The chemicals remained in contact with the cells for 1 hour at 30 °C and 150 rpm. Then the cells were washed with 8.5 g/L NaCl to reduce the concentration of the chemicals to a sub-inhibitory concentrations [16]. Adhered/biofilm cells were scraped and diluted in 8.5 g/L NaCl. Their cultivability was assessed in Mueller-Hinton Agar (MHA, Merck, Portugal). The number of colony forming units was determined after 24 hours at 30 °C incubation. Final results are presented as percentage of reduction compared with the situation without antimicrobial exposure.

2.4 Statistical analysis

Data were analysed applying the parametric paired t-test using the statistical program SPSS version 22.0 (Statistical Package for the Social Sciences). The average and standard deviation (SD) within samples were calculated for all cases. Statistical calculations were based on confidence level $\geq 95\%$ ($p < 0.05$) which was considered statistically significant. Three independent experiments were performed for each condition.

3. Results and Discussion

3.1 Antimicrobial activity of the chemicals

The increased demand for new products with antimicrobial properties has led to explore the wide variety of chemicals that constitute the phenolic group of phytochemicals. In this context two phytochemicals, cinnamic and ferulic acids were tested for their antimicrobial properties in comparison with two commonly used disinfectants, sodium hypochlorite and hydrogen peroxide. Initially, all the chemicals were tested for their antimicrobial efficiency. The MIC and MBC of the chemicals are presented in Table 1. Both the phytochemicals and the disinfectants were able to eradicate the bacteria. The most efficient chemical was sodium hypochlorite and both phytochemicals presented similar antimicrobial actions. Hydrogen peroxide was less efficient against *S. aureus* than on *E. coli*. This result can be due to the presence of catalase in *S. aureus*, an enzyme that degrades hydrogen peroxide into water and oxygen [18].

The antimicrobial properties demonstrated by sodium hypochlorite, hydrogen peroxide, cinnamic and ferulic acid are in agreement with the available literature where their effectiveness is highlighted, specially the disinfectants that are already widely used for disinfection purposes [19-24].

Table 1 Minimal inhibitory and bactericidal concentrations (MIC and MBC, respectively) of each chemical for *S. aureus* and *E. coli*

	S. aureus		*E. coli*	
	MIC (mM)	MBC (mM)	MIC (mM)	MBC (mM)
Sodium hypochlorite	4	5	3	3
Hydrogen peroxide	400	450	16	20
Cinnamic acid	25	25	15	> 25
Ferulic acid	25	> 25	> 25	> 25

3.2 Antiadhesive and antibiofilm activity

After assessing the efficacy of each chemical against planktonic *S. aureus* and *E. coli*, the effects of the selected antimicrobials was assessed on 2 hour adhered bacteria and 24 hour old biofilms (Fig. 1).

For most of the cases, adhered *E. coli* and their biofilms were more susceptible to the chemicals than *S. aureus*. This difference may be a consequence of the differences in membrane composition between the Gram-positive *S. aureus* and the Gram-negative *E. coli*. Contrarily to *S. aureus*, *E. coli* has an outer surface membrane composed by high lipid concentration and low peptidoglycan content [11] which may influence the antimicrobial action of the chemicals. In addition, *E. coli* ($p < 0.05$) was also more susceptible to hydrogen peroxide than *S. aureus* ($p > 0.05$), a result probably related with the presence of catalase.

Adhered cells exposed to sodium hypochlorite and cinnamic acid, were 100% inactivated on the polystyrene surfaces ($p < 0.05$). In fact, cinnamic acid was even more efficient than hydrogen peroxide against *S. aureus* (Fig. 1a).

Sodium hypochlorite is already known for its high antimicrobial efficiencies and broad spectrum of activity. The observed high effectiveness is mainly due to its mode of action once it is based on the concentration of hypochlorous acid and not on the total of free available chlorine concentration. Hypochlorous acid penetrates the bacteria across the cell wall and membranes. In addition, hypochlorous acid or hypochlorite inhibits essential enzymes activity that modulates growth, damages the membrane and DNA and is hypothesised that also injures membrane transport capacity [25, 26]. Nonetheless, the antimicrobial activity of sodium hypochlorite is dependent on the pH [25]. Furthermore, the considerable efficacy of the other disinfectant is also due to its mode of action once hydrogen peroxide produces destructive hydroxyl free radicals that are able to damage DNA that causes mutagenesis and which consequently kills bacteria and damage membrane, essential proteins and constituents. [27, 28]. Furthermore, it is known that hydrogen peroxide has bactericidal, virucidal, sporicidal and fungicidal properties. Nonetheless incorrect usage and storage leads to unstable hydrogen peroxide and loss of efficiency [20, 29].

Due to the promising results obtained for the phytochemicals, especially for cinnamic acid some additional assays were performed. Preliminary results demonstrate that cinnamic acid acts on bacterial surface diminishing its negative surface charge and hydrophilic character (data not shown). In fact, some authors observed that phenolic acids are weak organic acids (pKa $\approx$ 4.2) and their efficacy as antimicrobials depends on the concentration of undissociated acid. Once they present a lipophilic character, it is hypothesised that they cross the cell membrane by passive diffusion as undissociated chemical, disturb the cell membrane structure and probably acidify the cytoplasm and cause denaturation of proteins and increases bacteria permeability [30, 31]. In fact, that are evidences of ferulic acid to be capable of disrupting cell membranes which derail adhered *E. coli*. [30, 32, 33] The same effect was also accomplished for cinnamic acid against the yeast *Saccharomyces cerevisiae* and the bacterium *Listeria monocytogenes* [34, 35].

The chemicals were also capable to inactivate the bacteria in biofilms (Fig. 1b) despite not so pronounced as for the 2 h adhered cells. This different behaviour can be explained by the high resistance of biofilms [36]. However, cinnamic acid was capable to cause an inactivation of 40% ($p < 0.05$) whereas ferulic acid causes almost negligible inactivation. Sodium hypochlorite was almost 100% efficient against biofilm ($p < 0.05$). This chemical, despite its antibacterial efficiency has also an excellent cleaning action based on the synergism of the oxidizing power of hypochlorite and the capacity of hydroxide to dissolve organic soils [25, 26].

Phytochemicals have been demonstrating their advantage as a disinfectant to be used in several facilities, from industry to heath related areas, due to their antimicrobial efficacy, relatively low cost, generally low cytotoxicity and the fact that they are environmental friendly [37, 38].

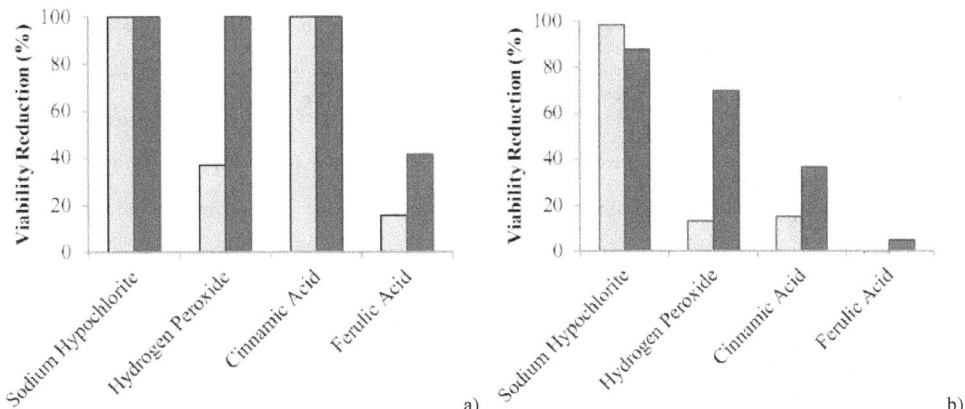

a)   b)

Fig. 1 The effects of the different chemicals on adhered cells (a) and biofilm (b) of *S. aureus* (□) and *E. coli* (■).

4. Conclusions

The overall results demonstrate that cinnamic acid has antimicrobial activity against *E. coli* and *S. aureus*. This phytochemical is able to effectively control monolayer adhered bacteria and reduce significantly the viability of biofilms, to values comparative to those obtained with sodium hypochlorite application. Therefore, cinnamic acid has antimicrobial, antiadhesive and a mild antibiofilm activity against the selected bacteria. Moreover, this study reinforces that phytochemicals are an auspicious alternative to commonly used disinfectants for general disinfection practices.

Acknowledgements The support of Operational Programme for Competitiveness Factors–COMPETE and by FCT–Portuguese Foundation for Science and Technology through Project Phytodisinfectants-PTDC/DTP-SAP/1078/2012 is gratefully acknowledged.

References

[1] Lopez GU, Kitajima M, Havas A, Gerba CP, Reynolds KA. Evaluation of a disinfectant wipe intervention on fomite-to-finger microbial transfer. Applied and Environmental Microbiology. 2014; 80:3113-8.

[2] Reichel M, Schlicht A, Ostermeyer C, Kampf G. Efficacy of surface disinfectant cleaners against emerging highly resistant gram-negative bacteria. BMC Infectious Diseases. 2014; 14:292-300.

[3] Kramer A, Schwebke I, Kampf G. How long do nosocomial pathogens persist on inanimate surfaces? A systematic review. BMC Infectious Diseases. 2006; 6:130-8.

[4] Hota B. Contamination, disinfection, and cross-colonization: Are hospital surfaces reservoirs for nosocomial infection? Clinical Infectious Diseases. 2004; 38:1182-9.

[5] Rutala WA, Weber DJ. New disinfection and sterilization methods. Emerging Infectious Diseases. 2001; 7:348-53.

[6] Chapman JS. Biocide resistance mechanisms. International Biodeterioration & Biodegradation. 2003; 51:133-8.

[7] Russel AD, Plasmids and bacterial resistance to biocides. Journal of Applied Microbiology. 1997; 82:155-65.

[8] Fraise AP. Biocide abuse and antimicrobial resistance – a cause for concern? Journal of Antimicrobial Chemotherapy. 2002; 49:11-2.

[9] Pimenta F, Abreu AC, Simões LC, Simões M. What should be considered in the treatment of bacterial infections by multi-drug therapies: A mathematical perspective? Drug Resistance Updates. 2014; 17:51-63.

[10] Borges A, Saavedra MJ, Simões M. The activity of ferulic and gallic acids in biofilm prevention and control of pathogenic bacteria. Biofouling. 2012; 28:755-67.

[11] Saavedra MJ, Borges A, Dias C, Aires A, Bennett RN, Rosa ES, Simões M. Antimicrobial activity of phenolics and glucosinolate hydrolysis products and their synergy with streptomycin against pathogenic bacteria. Journal of Medicinal Chemistry. 2010; 6:174-83.

[12] DeQueiroz GA, Day DF. Antimicrobial activity and effectiveness of a combination of sodium hypochlorite and hydrogen peroxide in killing and removing *Pseudomonas aeruginosa* biofilms from surfaces. Journal of Applied Microbiology. 2007; 103:794-802.

[13] Van Houdt R, Michiels CW. Biofilm formation and the food industry, a focus on the bacterial outer surface. Journal of Applied Microbiology. 2010; 109:1117-31.

[14] CLSI. Methods for Dilution Antimicrobial Susceptibility Tests for Bacteria that Grow Aerobically: Approved Standard - Ninth Edition M07- A09, In: NCCLS. 2012.

[15] Ferreira C, Pereira AM, Pereira MC, Melo LF, Simões M. Physiological changes induced by the quaternary ammonium compound benzyldimethyldodecylammonium chloride on *Pseudomonas fluorescens*. Journal of Antimicrobial Chemotherapy. 2011; 66:1036-43.

[16] Johnston MD, Lambert RJW, Hanlon GW, Denyer SP. A rapid method for assessing the suitability of quenching agents for individual biocides as well as combinations. Journal of Applied Microbiology. 2002; 92:784-9.

[17] Simões LC, Simões M, Oliveira R, Vieira MJ. Potential of the adhesion of bacteria isolated from drinking water to materials. Journal of Basic Microbiology. 2007; 47:174-83.

[18] Park B, Nizet V, Liu GY. Role of *Staphylococcus aureus* catalase in niche competition against *Streptococcus pneumoniae*. Journal of Bacteriology. 2008; 190:2275-8.

[19] Herald PJ, Davidson PM. Antibacterial activity of selected hydroxycinnamic acids. Journal of Food Science. 1983; 48:1378-9.

[20] McDonnell G, Russel AD. Antiseptics and disinfectants: activity, action and resistance. Clinical Microbiology Reviews. 1999; 12:147-79.

[21] Olasupo NA, Fitzgerald DJ, Gasson MJ, Narbad A. Activity of natural antimicrobial compounds against *Escherichia coli* and *Salmonella enterica* serovar Typhimurium. Letters in Applied Microbiology. 2003; 36:448-51.

[22] Pericone CD, Overweg K, Hermans PWM, Weiser JN. Inhibitory and bactericidal effects of hydrogen peroxide production by *Streptococcus pneumoniae* on other inhabitants. Infection and Immunity. 2000; 68:3990-7.

[23] Prasad VGNV, Swamy PL, Rao TS, Rao GS. Antibacterial synergy between oxytetracycline and selected polyphenols against bacterial fish pathogens. International Journal of Veterinary Science. 2013; 2:68-70.

[24] Rutala WA, Weber DJ. Infection control: the role of disinfection and sterilization. Journal of Hospital Infection. 1999; 43:S43-S55.

[25] Estrela C, Estrela CRA, Bardin EL, Spanó JC, Marchesan MA, Pécora JD. Mechanisms of action of sodium hypochlorite. Brazilian Dental Journal. 2002; 13:113-7.

[26] Fukuzaki S. Mechanisms of action of sodium hypochlorite in cleaning and disinfection processes. Biocontrol Science. 2006; 4:147-57.

[27] Imlay JA, ChinSM, Linn S. Toxic DNA damage by hydrogen peroxide through the fenton reaction *in vivo* and *in vitro*. 1988; 240:640-2.

[28] Storz G, Tartaglia LA, Ames BN. Bacterial defenses against oxidative stress. Trends in Genetics. 1990; 6:363-8.

[29] Mitiku M, Ali S, Kibru G. Antimicrobial drug resistance and disinfectants susceptibility of *Pseudomonas aeruginosa* isolates from clinical and environmental samples in Jimma University Specialized Hospital, Southwest Ethiopia. 2014; 2:40-5.

[30] Campos FM, Couto JA, Figueiredo AR, Tóth IV, Rangel AOSS, Hogg TA. Cell membrane damage induced b phenolic acids on wine lactic acid bacteria. 2009; 135:144-51.

[31] Johnston MD, Hanlon GW, Denyer SP, Lambert RJW. Membrane damage to bacteria caused by single and combined biocides. Journal of Applied Microbiology. 2003; 94:1015-23.

[32] Borges A, Ferreira C, Saavedra MJ, Simões M. Antibacterial activity and mode of action of ferulic and gallic acids against pathogenic bacteria. Microbial Drug Resistance. 2013; 19:256-65.

[33] Lou Z, Wang H, Rao S, Sun J, Ma C, Li J. *p*-Coumaric acid kills bacteria through dual damage mechanisms. Food Control. 2012; 25:550-4.

[34] Chambel A, Viegas CA, Sá-Correia I. Effect of cinnamic acid on the growth and on plasma membrane H^+-ATPase activity of *Saccharomyces cerevisiae*. International Journal of Food Microbiology. 1999; 50:173-9.

[35] Ramos-Nino ME, Clifford MN, Adams MR. Quantitative structure activity relationship for the effect of benzoic acids, cinnamic acids and benzaldehydes on *Listeria monocytogenes*. Journal od Applied Bacteriology. 1996; 80:303-10.

[36] Davies D. Understanding biofilm resistance to antibacterial agents. Nature Reviews Drug Discovery. 2003; 2:114-22.

[37] Abreu AC, McBain AJ, Simões M. Plants as sources of new antimicrobials and resistance-modifying agents. Natural Products Report. 2012; 29:1007-21.

[38] Ou S, Kwok K-C. Ferulic acid: pharmaceutical functions, preparation and applications in foods. Journal of the Science of Food and Agriculture. 2004; 84:1261-9.

Decreased susceptibilities to biocides and resistance to antibiotics following adaptation to quaternary ammonium compounds in food-associated bacteria

C. Soumet[*,1], D. Méheust[2], C. Pissavin[2], P. Le Grandois[1], B. Frémaux[3], C. Feurer[3], A. Le Roux[3], M. Denis[4] and P. Maris[1]

[1] ANSES, Antibiotics, Biocides, Residues and Resistance Unit, Fougères Laboratory, France
[2] IUT Saint-Brieuc, Biology Department, University of Rennes 1, France
[3] IFIP, Department of Fresh and Processed Meat, Maisons-Alfort and Le Rheu, France
[4] ANSES, Hygiene and Quality of Poultry and Pork Products Unit, Ploufragan-Plouzané Laboratory, France
*Corresponding author: e-mail: christophe.soumet@anses.fr, Phone: +33 299947857

Appropriate use of biocides is crucial to control and prevent pathogenic bacteria along the food chain. However, in some circumstances, bacteria can be exposed to sub-inhibitory concentrations of biocide exerting a selective pressure that would lead to the selection for bacteria less susceptible to biocides and/or antibiotics. The aim of this study was to assess the effects of repeated exposure to didecyl dimethyl ammonium chloride (DDAC) on the susceptibility to biocides and antibiotics in *Campylobacter coli*, *Salmonella enterica*, *Listeria monocytogenes* and *Escherichia coli*. Following adaptation to DDAC, decreased susceptibilities to other quaternary ammonium compounds and to various antibiotics were observed in the different species. Most of the *E. coli* strains acquired resistances to antibiotics critical for human medicine. Thus, extensive use of DDAC at sub-inhibitory concentrations may lead to the development of antibiotic-resistant bacteria and may represent a public health issue.

Keywords Adaptation, Disinfectants, Antibiotics, Decreased susceptibility, Resistance, QAC

1. Introduction

Biocides are chemical agents playing a crucial role in limiting the spread of infections and diseases. They are widely used in different fields (human medicine, agriculture, household field...). In the modern food industry, the use of biocides has considerably increased to satisfy consumer demands of healthy, nutritious, and minimally processed foods and to ensure food safety. The quaternary ammonium compounds, such as benzalkonium chloride (BC) and didecyl dimethyl ammonium chloride (DDAC), are among the most widely used biocides in cleaning and disinfection steps. However, there are growing concerns about the risk of selection of resistant bacteria to antibiotics with regard to the increased use of biocides.

It is well acknowledged that the main reasons for the increasing prevalence of antibiotic resistances are the widespread usage of antibiotics in human and veterinary therapy and the inclusion of antibiotics in animal feedstuffs [1]. The contribution of other factors such as disinfectant usage is nevertheless difficult to evaluate.

Disinfectants as biocides are generally effective to inhibit or kill bacteria when they are applied at the concentrations recommended by the manufacturer. But, they can be found at lower concentrations in contact with bacteria because of insufficient cleaning or rinsing before disinfection, under-dosing of applied disinfectant, biofilm formation ... Under such conditions, bacteria are exposed to sub-lethal concentrations of disinfectants, and this can lead to adaptation of initially susceptible bacteria [2].

Furthermore, repeated exposure to sub-inhibitory concentrations of biocide may promote the decreased susceptibility or resistance to other antimicrobial agents. Bacteria may indeed elicit common cellular responses to counteract the effects of biocides and antibiotic [3]: reduced permeability or uptake, enhanced efflux, enzymatic inactivation... This phenomenon has been known as cross-susceptibility or cross-resistance between biocides and antibiotics, but the mechanisms are still poorly understood.

Although the adaptation to sub-inhibitory concentrations of biocide has been documented for some bacteria [4-7], the number of bacterial strains studied is usually limited. The aim of the present study was to assess the effects of step-wise exposure to DDAC on the antimicrobial susceptibilities of 136 food-associated bacterial strains. The analysis included the susceptibility to the biocide itself, the susceptibility to 4 other disinfectants and the cross-resistance to clinically important antimicrobials. Four species were chosen in this study (*Campylobacter coli, Salmonella enterica, Listeria monocytogenes* and *Escherichia coli*) as they belong to the nine biological hazards transmitted to human through the consumption of pork products.

2. Materials and Methods

2.1   Bacterial strains

Strains isolated from pig production chain were studied for each of the following species: *Escherichia coli* (n=54), *Listeria monocytogenes* (n=31), *Salmonella enterica* (serotype Typhimurium and Derby) (n=35) and *Campylobacter coli* (n=16). All the strains (n=136) were kept at -80°C in a nutritive solution containing 20% (v/v) glycerol.

2.2   Biocide susceptibility testing

The disinfectants used in this study are commonly found in biocide formulations used in the French pork industry. They included didecyl dimethyl ammonium chloride (DDAC), benzalkonium chloride (BC), sodium hypochlorite (NaOCl) and hydrogen peroxide (H_2O_2). A commercially available food-grade biocide formulation (Galox Horizon) was also used, containing DDAC and glutaraldehyde. Susceptibility tests against the 4 active substances and the commercial biocide formulation were performed using a standard microdilution method [8].

2.3   Antibiotic susceptibility testing

Antibiotic susceptibility tests were performed using a standard microdilution method with the Sensititre® system on customized microtiter plates (Trek Diagnostic Systems, England). The strains were interpreted as susceptible or resistant to antibiotics according to the epidemiological resistance cut-off determined from EUCAST (European Committee on Antimicrobial Susceptibility Testing) and CA-SFM (Antibiogram Committee of the French Society for Microbiology) websites. The number of tested antibiotics differed depending on the species: 20 antibiotics for *L. monocytogenes*, 14 for *S. enterica*, 7 for *C. coli* and 13 for *E. coli*.

2.4   Adaptation experiments to DDAC, a quaternary ammonium compound

Adaptive responses of bacterial strains were investigated by exposing the strains daily to increasing sub-inhibitory concentrations of DDAC for 7 days, as described previously [9]. Briefly, a calibrated bacterial suspension of 10^8 Colony-Forming Unit (CFU)/ml (100 μl) was initially exposed to a starting concentration of disinfectant below the Minimum Inhibitory Concentration (MIC) for 24 h at 37 °C in a total volume of 10 ml of Mueller-Hinton (MH) broth. When growth was observed, a 10-fold diluted culture was transferred to fresh MH broth supplemented with a higher concentration of disinfectant. If no growth was observed, the previous concentration was used. As control, a bacterial suspension (100 μl) and MH broth (10 ml) without disinfectant were tested using the same protocol. After 7 days, bacteria were spread with a loop (10 μl) on MH agar and incubated for 24 h at 37 °C. They were then collected with 2.5 ml of storage nutritive solution and kept in cryotubes at −80 °C. MIC increase factor for each biocide and each antibiotic was determined for each strain as a ratio between MIC after adaptation and MIC before adaptation.

3. Results

3.1   Reduced susceptibility to biocides

Before the adaptation experiment, DDAC MIC values varied from 0.75 to 1.5 μg/ml among *Listeria* strains, from 2 to 4 μg/ml among *Salmonella* strains, from 0.37 to 0.75 μg/ml among *Campylobacter* strains and from 2 to 4 μg/ml among *E. coli* strains. Following adaptation to DDAC, a 3-fold increase in the MIC values for this biocide was observed in about 50% of the *E. coli* and *L. monocytogenes* strains, and 10% of the *Salmonella* strains (Table 1). The increase was lower (2-fold) in *Campylobacter coli*.

Table 1   Biocide MIC increase factor for different bacteria species after exposure to increasing concentrations of DDAC.

Species	DDAC	BC	NaOCl	H_2O_2	Galox Horizon
Escherichia coli (54)[a]	≥3 (48%)[b]	≥3 (22%)	<2 (100%)	<2 (100%)	≥3 (65%)
Listeria monocytogenes (31)	≥3 (48%)	≥3 (45%)	<2 (100%)	<2 (100%)	≥3 (45%)
Salmonella enterica (35)	≥3 (10%)	≥1.5 (40%)	<1.5 (100%)	<1.5 (100%)	≥3 (6%)
Campylobacter coli (16)	2 (31%)	3 (31%)	<2 (100%)	<2 (100%)	3 (31%)

DDAC: didecyl dimethyl ammonium chloride; BC: benzalkonium chloride; NaOCl: sodium hypochlorite;
H_2O_2: hydrogen peroxide; Galox Horizon: biocide formulation containing DDAC and glutaraldehyde
[a]: number of strains tested for a given species; [b]: percentage of strains with the biocide MIC increase factor indicated

Reduced susceptibility to other biocides was found with the most important increase in MIC for BC and the commercial biocide formulation, for all species except *Salmonella*. No significant difference was observed in the susceptibility to hydrogen peroxide and sodium hypochlorite for all strains tested.

3.2 Reduced susceptibility and resistance to antibiotics

Following adaptation to DDAC, increase in antibiotic MIC values was more pronounced in *E. coli* in terms of antibiotics numbers and of magnitude (from 4- to 32-fold increase) and, to a lesser extent, in *Salmonella* strains. The evolution of antibiotic susceptibility was low in *Listeria monocytogenes* and *Campylobacter coli*: MICs increased from 4- to 8-fold for a few antibiotics.

Fifty of the 54 strains of *E. coli* were resistant to at least one antibiotic after adaptation, meaning that their MIC values became superior to the epidemiological cut-offs. Most of them acquired resistance to 1 or 2 new antibiotics, and six strains became multiresistant (defined as resistant to at least five antibiotics). Most of these strains had acquired resistance to ampicillin, cefotaxime, ceftazidime, chloramphenicol and ciprofloxacin (Fig. 1). These results suggest that nonspecific mechanisms, such as efflux pumps, may be responsible for this multiresistance.

Some *Salmonella* strains acquired resistance to various antibiotics: chloramphenicol (n=3 strains), florfenicol (n=1), nalidixic acid (n=1), ciprofloxacin (n=2) and streptomycin (n=2). Only one strain of *Listeria monocytogenes* and two strains of *Campylobacter coli* became resistant to tetracycline and streptomycin, respectively.

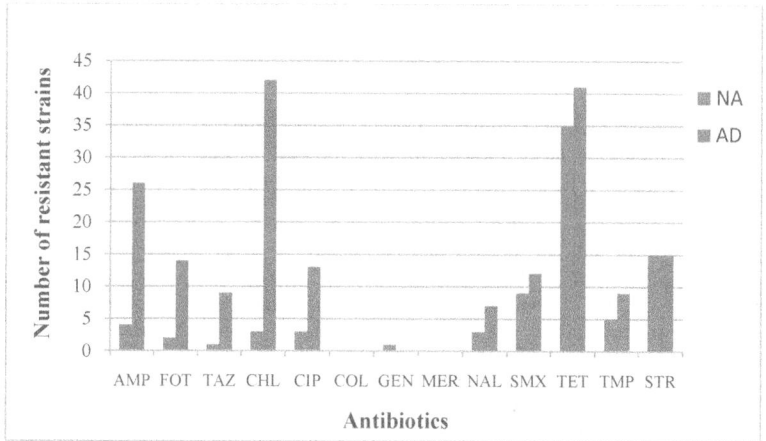

Fig. 1 Number of *E. coli* strains resistant to antibiotics before and after adaptation to DDAC.
NA: before adaptation test; AD: after adaptation test
AMP: ampicillin; FOT: cefotaxime; TAZ: ceftazidime; CHL: chloramphenicol; CIP: ciprofloxacin; COL: colistin; GEN: gentamicin; MER: meropenem; NAL: nalidixic acid; SMX: sulfamethoxazole; TET: tetracyclin; TMP: trimethoprim; STR: streptomycin.

4. Discussion

4.1 Adaptation to DDAC and decreased susceptibility to other biocides

Adaptation to biocides by serial passage in increasing sub-inhibitory concentrations has been documented for some bacterial species. Similarly to antibiotics, the word "resistance" has been widely employed also for biocides used as disinfectants. There is however no clear consensus in the scientific community on the definition of biocide resistance, because available epidemiological data in this respect is extremely limited [10, 11]. The term "decreased (or reduced) susceptibility" was preferred in the present paper and the MIC increase factor was used for describing adaptive response to disinfectant exposure.

The ability of food-associated bacteria to survive in increasing concentrations of quaternary ammonium compounds has been investigated, but mainly by using benzalkonium chloride (BC). In this work, the effect of a repeated exposure to didecyl dimethyl ammonium chloride (DDAC) was studied on four pathogen species, because this biocide compound is frequently used in food industry.

Following adaptation, a low increase in the DDAC MIC values was observed in *Campylobacter coli* and *Salmonella* strains. Furthermore, no significant difference in the susceptibility to other biocides was found in *Salmonella* strains. These results are consistent with a previous study, where no stable decrease in the biocide susceptibility was detected in *Salmonella* after several rounds of *in vitro* selection by increasing concentrations of active biocides [7]. Concerning *C. coli*, Mavri and coworkers showed that a statistically significant difference was observed between one strain exposed to BC and the control strain after 15 days of exposure [6]. The period of exposure in the current study may not be sufficient to observe significant biocide susceptibility changes in *C. coli*, and this test was carried out on a few strains.

A 3-fold increase in the DDAC MIC values was observed in about half of the *E. coli* and *L. monocytogenes* strains tested. Using the same adaptation experiment, Soumet and coworkers formerly found similar levels of reduction in susceptibility to different QACs in *E. coli* (with a mean 3-fold increase) [9]. Previous studies showed that the reduced susceptibility to BC varied from 10% to 46% in *L. monocytogenes* isolated from foods and food processing environment [8, 12, 13] and that adaptation to various disinfectants is common in this species [5]. From the strong adaptation ability observed here, it can be suspected that the prevalence of decreased susceptibility to DDAC in *Listeria* is relatively high in food industry.

The impact of DDAC exposure on reduced susceptibility to other biocides depended on the biocide type tested. Reduced susceptibilities were more pronounced for QAC disinfectants (BC and the commercial biocide formulation containing DDAC). Whatever the species, adaptation to DDAC seemed to not induce reduced susceptibilities to peroxygens (H_2O_2) and chlorine releasing agents (NaOCl).

4.2   Effect of DDAC adaptation on reduced susceptibility and resistance to antibiotics

A growing concern is the possibility that mechanisms providing reduced susceptibility to biocides may also provide cross-protection to clinically important antibiotics [14]. When resistance is expressed, strains can move from the category of "sensitive" to that of "resistant" according to the CLSI (Clinical Laboratory Standard Institute) guidelines on antimicrobial susceptibility testing.

In this study, MICs increased by 4-fold for erythromycin and ciprofloxacin in two *C. coli* strains following adaptation to DDAC. A previous work found that the resistance of ciprofloxacin increased only by 2-fold after 5 passages in presence of BC for one *C. coli* strain [6]. Furthermore, Mavri and coworkers showed that in 29% of cases, after step-wise exposure to biocides, the *Campylobacter* strains were indeed more susceptible to the tested antimicrobials than the parent and the control strains.

The evolution of antibiotic susceptibility was low in *L. monocytogenes* strains in the present work (with only one strain becoming resistant to tetracycline); that is consistent with a previous study where no resistance to antibiotics was found in this species [8]. Compared to *L. monocytogenes*, increases in antibiotic MIC values were more pronounced in *Salmonella* strains (with the highest increase of 32-fold for nalidixic acid in one strain). Resistances to antibiotics were nevertheless acquired for only seven out of 35 *Salmonella* strains. Braoudaki and Hilton also found a low degree of resistance to a range of antibiotics in *Salmonella* serovar Typhimurium following adaptation to BC [15]. Reduced susceptibility to several antibiotics was also observed after exposure of *Salmonella* Typhimurium strains to a biocidal formulation containing a QAC [16].

The impact of DDAC exposure on reduced susceptibility was stronger in *E. coli* than in the other species, in terms of antibiotics numbers and of magnitude. The highest MIC increase factor was of 32-fold for chloramphenicol, and 72% of the tested strains having acquired the resistance to this antibiotic. Similar results were found in a previous study where 90% of *E. coli* strains became resistant to phenicol compounds following adaptation to different QACs [9]. Other classes of antibiotic showing a marked increase in resistance were penicillins (ampicillin), aminoglycosides (streptomycin), cephalosporins (cefotaxime and ceftazidime) and fluoroquinolones (ciprofloxacin and nalidixic acid). The three last classes were also identified by Capita and colleagues, after exposure of *E. coli* ATCC 12806 to sublethal concentrations of food-grade biocides [17]. It is noteworthy that the study presented here revealed an increase in resistance to quinolones after adaptation to DDAC, because fluoroquinolones are critically important drugs for treating serious *E. coli* infections in humans. Furthermore, the movement of strains from the category of "sensitive" to that of "resistant" observed for aminoglycosides and cephalosporins is also a matter of concern, as these compounds are the front-line antimicrobials for treating serious bacterial infections.

In conclusion, adaptation to DDAC differed depending on the species studied. Exposure to sub-inhibitory QAC concentrations could lead to the development of antibiotic-resistance bacteria, especially in *E. coli*. However, many authors reported that changes in the pattern of susceptibility to antibiotics and biocides might be strain-specific rather than species-specific. Adaptation mechanism to biocide was indeed found to be unique for each strain of *Campylobacter* [6], and reduced susceptibility to antimicrobials seemed to differ depending on the serotype of *Listeria monocytogenes* and *Salmonella enterica* [7, 12, 13, 18]. These intra-specific differences underline the need to screen a wide range of test strains as mentioned by Maillard and coworkers [11].

Acknowledgements DABESBIO project is funded by the Agriculture Ministry in the framework of the contract « Plan Etat – Région Bretagne ».

References

[1]   Ortega Morente E., Fernández-Fuentes M.A., Grande Burgos M.J., Abriouel H., Pérez Pulido R.,Gálvez A. Biocide tolerance in bacteria. International Journal of Food Microbiology. 2013; 162(1): p. 13-25.

[2]   Davidson P.M.,Harrison M.A. Resistance and adaptation to food antimicrobials, sanitizers, and other process controls. Food technology. 2002; 56: 69-78.

[3]   Buffet-Bataillon S., Branger B., Cormier M., Bonnaure-Mallet M.,Jolivet-Gougeon A. Effect of higher minimum inhibitory concentrations of quaternary ammonium compounds in clinical *E. coli* isolates on antibiotic susceptibilities and clinical outcomes. Journal of hospital infection. 2011; 79: 141-146.

[4]   Braoudaki M.,Hilton A.C. Mechanisms of resistance in *Salmonella enterica* adapted to erythromycin, benzalkonium chloride and triclosan. International Journal of Antimicrobial Agents. 2005; 25: 31-7.

[5]   Lunden J., Autio T., Markkula A., Hellstrom S.,Korkeala H. Adaptive and cross-adaptive responses of persistent and non-persistent *Listeria monocytogenes* strains to disinfectants. International Journal of Food Microbiology. 2003; 82: 265-272.

[6]   Mavri A.,Mozina S.S. Development of antimicrobial resistance in *Campylobacter jejuni* and *Campylobacter coli* adapted to biocides International Journal of Food Microbiology. 2013 160: 304-312

[7]   Condell O., Iversen C., Cooney S., Power K.A., Walsh C., Burgess C.,Fanning S. Efficacy of biocides used in the modern food industry to control *Salmonella enterica*, and links between biocide tolerance and resistance to clinically relevant antimicrobial compounds. Applied and Environmental Microbiology. 2012; 78: 3087-3097.

[8]   Soumet C., Ragimbeau C.,Maris P. Screening of benzalkonium chloride resistance in *Listeria monocytogenes* strains isolated during cold smoked fish production. Letters in Applied Microbiology. 2005; 41: 291-6.

[9]   Soumet C., Fourreau E., Legrandois P.,Maris P. Resistance to phenicol compounds following adaptation to quaternary ammonium compounds in *Escherichia coli*. Veterinary Microbiology. 2012; 158: 147-152.

[10] Morrissey I., Oggioni M.R., Knight D., Curiao T., Coque T., Kalkanci A., Martinez J.L., Baldassarri L., Orefici G., Yetiş Ü., Rödger H.J., Visa P., Mora D., Leib S.,Viti C. Evaluation of epidemiological cut-off values indicates that biocide resistant subpopulations are uncommon in natural isolates of clinically-relevant microorganisms. PLOS one. 2014; 9.

[11] Maillard J.Y., Bloomfield S., Coelho J.R., Collier P., Cookson B., Fanning S., Hill A., Hartemann P., Mcbain A., Oggion I.M., Sattar S., Schweizer H.,Threlfall J. Does microbicide use in consumer products promote antimicrobial resistance? A critical review and recommendations for a cohesive approach to risk assessment. Microbial drug resistance. 2013; 19:344-354.

[12] Xu D., Li Y., Shamim Hasan Zahid M., Yamasaki S., Shi L., Li J.-R.,Yan H. Benzalkonium chloride and heavy-metal tolerance in *Listeria monocytogenes* from retail foods. International Journal of Food Microbiology. 2014; 190: 24-30.

[13] Aase B., Sundheim G., Langsrud S.,Rorvik L.M. Occurrence of and a possible mechanism for resistance to a quaternary ammonium compound in *Listeria monocytogenes*. International Journal of Food Microbiology. 2000; 62: 57-63.

[14] Russell A.D. Mechanisms of antimicrobial action of antiseptics and disinfectants : an increasingly important area of investigation Journal of antimicrobial chemotherapy. 2002; 49:597-599.

[15] Braoudaki M.,Hilton A.C. Adaptive resistance to biocides in *Salmonella enterica* and *Escherichia coli* O157 and cross-resistance to antimicrobial agents. Journal of Clinical Microbiology. 2004; 42:73-78.

[16] Karatzas K.A., Webber M.A., Jorgensen F., Woodward M.J., Piddock L.J.,Humphrey T.J. Prolonged treatment of *Salmonella enterica* serovar Typhimurium with commercial disinfectants selects for multiple antibiotic resistance, increased efflux and reduced invasiveness. Journal of Antimicrobrial Chemotherapy. 2007; 60: 947-55.

[17] Capita R., Riesco-Peláez F., Alonso-Hernando A.,Alonso-Calleja C. Exposure of *Escherichia coli* ATCC 12806 to Sublethal Concentrations of Food-Grade Biocides Influences Its Ability To Form Biofilm, Resistance to Antimicrobials, and Ultrastructure. Applied and Environmental Microbiology. 2014; 80: 1268-1280.

[18] Braoudaki M.,Hilton A.C. Adaptive resistance to biocides in *Salmonella enterica* and *Escherichia coli* O157 and cross-resistance to antimicrobial agents. Journal of Clinical Microbiology. 2004; 42:73-78.

Effect of storage duration on microbial load of orange pomace

A.O. Oduntan[1,*], O.B. Fajinmi[2] and O.E. Oyedeji[3]

[1] Product development programme, National Horticultural Research Institute, P.M.B. 5432, Idi-Ishin, Ibadan, Oyo State, Nigeria.
[2] Fruit Research Programme National Horticultural Research Institute, P.M.B. 5432, Idi-Ishin, Ibadan, Oyo State, Nigeria.
[3] Citrus Research Programme, National Horticultural Research Institute, P.M.B. 5432, Idi-Ishin, Ibadan, Oyo State, Nigeria.
*Corresponding e-mail bosetunde12@yahoo.com

Residue from the processing of fruits and vegetables, traditionally considered as an environmental problem are being increasingly recognized as sources for obtaining high-phenolic products. The objective of this work is to determine the extent of microbial load of orange pomace with storage duration in order to establish its safety for new products for human consumption. Orange pomace was collected after juice extraction from the Product Development Programme of the National Horticultural Research Institute, Ibadan, Nigeria. Pomace samples were kept at ambient temperature of $32^{O}C$, representative samples were taken at every two hours over a period of eight hours for microbial analysis. The culture media used were Potato Dextrose Agar (PDA) to enumerate fungi and Nutrient Agar (NA) to enumerate bacteria. The result showed increase in total bacterial load from 0hr to 8hr, the highest count was 8hr with total count of 2.6×10^{3} cfu/g while the least was at 0hr with total count of 1.1×10^{2} cfu. The difference in the count was statistically significant at $p < 0.5$. After 48hr of incubation, there was decline in total fungal from 0hr (5.4×10^{2}) to 8hr (0 cfu/g). Significant variation was observed in the total plate count at $p < 0.05$, Progressive decrease was observed in *Aspergillus flavus* count from 0hr to 8hr of sampling while there was decrease in the total *Aspergillus niger* from 0hr to 8hr except at 6hr, where sudden increase in microbial load was observed. Orange pomace can be used for new products not more than 6hrs after juice extraction to avoid microbial contamination.

Keywords: Orange pomace; storage duration; microbial load; contamination

Introduction

Fruit and vegetable juices have become important in recent years due to overall increase in natural juice consumption as an alternative to the traditional caffeine containing beverages such as coffee, tea, or carbonated soft drinks [1]. Residue from the processing of fruits and vegetables, traditionally considered as an environmental problem are being increasingly recognized as sources for obtaining high-phenolic products. The polyphenolics from waste materials, being derived from agro-industrial production, may be used as functional food ingredients and as natural antioxidants [2].

Fruit and vegetable wastes are inexpensive, available in large quantities, characterized by a high dietary fiber content resulting with high water binding capacity and relatively low enzyme digestible organic matter [3]. Due to the high dietary fiber content and contrasting dietary fiber properties, the coproducts could be used to change physicochemical properties of diets.

There is a huge amount of by products in food industries that process vegetable and fruits. [4] Citrus peel, remaining after juice extraction, is the primary waste fraction amounting to almost 50% of the fruit mass [5]. Pineapple pomace constitutes 35% of waste generated in pineapple juice production.

The processing of fruits results in high amounts of waste materials such as peels, seeds, stones, and oilseed meals which constitute environmental problem. A disposal of these materials usually represents a problem that is further aggravated by legal restrictions. Thus new aspects concerning the use of these wastes as by-products for further exploitation on the production of food additives or supplements with high nutritional value have gained increasing interest because these are high-value products and their recovery may be economically attractive.

A number of researchers have used fruits and vegetable by-products such as apple, pear, orange, peach, blackcurrant, cherry, artichoke, asparagus, onion, carrot pomace [6], [7], [8] as sources of dietary fibre supplements in refined food.

It is well known that by-products represent an important source of sugars, minerals, organic acid, dietary fibre and phenolics which have a wide range of action which includes antitumoral, antiviral, antibacterial, cardioprotective and antimutagenic activities [9]. The objective of this work was to determine the extent of microbial load of Orange pomace with time in order to establish its safety for human consumption.

Methodology

Orange pomace was collected after juice extraction from the Product Development Programme of the National Horticultural Research Institute, Ibadan, Nigeria. Pomace samples were kept at ambient temperature of $32^{O}C$, representative samples were taken at every two hours over a period of eight hours for microbial analysis.
Sample preparation: samples were milled with a blender model Marlex Excella, 10g of orange pomace each were diluted with 90ml of sterile distil water and shaken with orbital shaker model Gallenkamp for 2mins. 1mL was used to prepare a ten fold serial dilution to obtain a dilution range of $10^{-1} - 10^{-6}$ for each sample, 1mL of 10^{-1}, 10^{-3}, 10^{-6} dilution were pipetted into 9cm diameter petri dish, 15mL of molten sterilized culture medium cooled to about $45^{O}C$ was poured into each plate and gently swirled to homogenize the mixture. The culture media used were Potato Dextrose Agar (PDA) to enumerate fungi and Nutrient Agar (NA) to enumerate bacteria. Petri dishes containing NA were incubated at $27^{O}C$ for 24hrs while those with PDA were incubated at $27^{O}C$ for 48hrs before enumeration. Colony counter of model Lapiz digital was used for the enumeration.

Result and Discussion

Progressive increase in total bacteria count was observed from samples collected from 0 to 8hr, the highest count was 8hrs with total count of 2.6×10^{3}cfu while the least was at 0hr with total count of 1.1×10^{2} cfu. The difference in the count was statistically significant {0hr (119), 2hr (102, 4hr (144), 6hr (177) and 8hr (2598)]. From the result, a very high rapid increase in bacteria count was observed from 6hr to 8hr, this suggest that the pomace can be utilize not more than 6hrs after juice extraction at ambient temperature of $32^{O}C$.

After 48hr of incubation, there was decline in total fungal from 0hr to 8hr as shown in Plate 1. Significant variation was observed in the total plate count at $p < 0.05$. Significant difference was observed in the total fungal count at 0 (535), 6 (344). At 8hrs (0) total inhibition of fungal growth was observed. This could be attributed to release of phytotoxic substances from the pomace which has inhibitory effect on the growth of micro-organism like fungi [10].

Progressive decrease was observed in the *Aspergillus flavus* load count throughout the sampling period. The growth of *A flavus* was totally inhibited at 8hr which suggests that phytotoxic substances produced by the pomace prevented growth of *A. flavus* which eventually led to the death of *A flavus* that initially grew on the pomace. Significant differences were observed in 0, 2, and 8hr at $p < 0.05$.

Table 1: Effect of storage duration on bacterial and fungal load in Orange pomace

Bacterial Load Storage hour	Total count	Fungal load Total count	A. Flavus	A. niger
0	1.1×10^{2d}	534.7^{a}	65.33^{a}	9.33^{b}
2	1.2×10^{2d}	338.7^{ab}	25.00^{b}	1.33^{c}
4	1.4×10^{2c}	261.3^{b}	8.0^{bc}	3.33^{c}
6	1.8×10^{2b}	344.0^{ab}	6.67^{bc}	37.0^{a}
8	2.6×10^{3a}	0.0^{c}	0.0^{c}	0.0^{c}

Generally, there was decrease in the total *Aspergillus niger* from 0hr to 8hr except at 6hr, where sudden increase in microbial load was observed. There was no significant difference in the *Aspergillus niger* growth at 2, 4 and 8hr while significant difference was observed between 0hr and 6hr which was significantly higher than 2, 4 and 8hr.

Conclusion

A very high rapid increase in bacteria count was observed from 6hr to 8hr, this suggest that the pomace cannot be utilized more than 6hrs after juice extraction to avoid bacterial contamination. On the contrary, as the storage time increases, there was decrease in fungal load, this implies that orange pomace could be utilized after 8hr without fungal contamination.

Acknowledgement The support of Society of General Microbiology (SGM) and African Women in Agricultural Research and Development (AWARD) is greatly acknowledged.

References

[1] Kaur S, Sarkar BC, Sharma HK, Singh C. Response surface optimization of conditions for the clarification of guava fruit juice using commercial enzyme. Journal of Food Processing and Engineering. 2009; 10.1111/j, 1745- 4530.

[2] Zhou S, Fang Z, Lu Y, Chen J, Liu D & Ye X. Phenolics and antioxidant properties of bayberry (*Myrica rubra* Sieb. et Zucc.) pomace. Food Chemistry. 2009; 112: 394- 399

[3] Serena A. & Kundsen B. Chemical and physicochemical characterisation of co-products from vegetable food and agro industries, Animal Feed Science and Technology. 2007; 139:109–124.

[4] Bocco A, Cuvelier MA, Richard Hand Berset C. Antioxidant activity and phenolic composition of citrus peel and seed extracts. Journal of Agricultural and Food Chemistry. 1998; 46:2123–2129.

[5] Braddock RJ, By – products of citrus fruits. Food Technology.1995; 49: 76-77.

[6] Grigelmo-Miguel N. and Martin-Belloso O. (1999) Influence of fruit dietary fiber addition on physical and sensorial properties of strawberry jams. Journal of Food Engineering. 1999; 41: 13 – 21.

[7] Ng A., Lecain S., Parker M.L., Smith A.C., Waldron K.W. Modification of cell wall polymers of onion waste lll. Effect of extrusion-cooking on cell wall material of outer fleshy tissue. Carbohydrate polymers.1999; 39: 341 – 349

[8] Nawirska A., Kwasniewska M. Dietary fibre fractions from fruit and vegetable processing waste. Food Chemistry.2005; 91(29): 221 – 225

[9] Djilas, S., Čanadanović-Brunet, J. and Ćetković, G. By-Products of Fruits Processing as a source of Phytochemicals. Chemical Industry and Chemical Engineering Quarterly. 2009; 15(4) 191 – 202.

[10] Nigam PS, A. Pandey (eds.), Biotechnological Potential of fruit Processing Industry Residues. Biotechnology for Agro-Industrial Residues Utilisation. C _ Springer Science+Business Media B.V; 2009. P. 276-286.

Effectiveness of phage-based probiotic dietary supplement in the prevention of E.coli traveler's diarrhea: a small-scale study

A. V. Aleshkin[*,1,2], **E. O. Rubalskii**[1,2], **N. V. Volozhantsev**[3], **V. V. Verevkin**[3], **E. A. Svetoch**[3], **I. A. Kiseleva**[1,2], **S. S. Bochkareva**[2], **O. Yu. Borisova**[1,2], **A. V. Popova**[1,3], **A. G. Bogun**[3] and **S. S. Afanas'ev**[2]

[1]Bphage LLC, 10 Butirskiy val str., 125047 Moscow, Russia
[2]Gabrichevsky Moscow Research Institute for Epidemiology and Microbiology, 10 Admirala Makarova str., 125212 Moscow, Russia
[3]State Research Center for Applied Microbiology & Biotechnology, 142279 Obolensk, Moscow region, Russia
*Corresponding author: e-mail: ava@gabri.ru, Phone: +7 9853341032

Traveler's diarrhoea (TD) is caused by Escherichia coli in 30% of cases. We have developed a phage cocktail for prophylaxis of TD caused by E.coli, Shigella flexneri, Shigella sonnei, Salmonella enterica, Listeria monocytogenes or Staphylococcus aureus, and investigated its effectiveness against infection caused by the non-pathogenic Lac (-) strain of E.coli K12 C600 in animal and human trials. On the 6th day of both animal and human trials E. coli K12 C600 strain was detected in titre of 10^4 CFU/g of mice feces and 10^6 CFU/g of human feces in the control (untreated) groups, while it was not detected in the samples of either of the study (phage-treated) groups. These results have great significance because the original coliphages included in the cocktail have a broad host-range including ETEC, EAEC and EHEC strains which cause severe cases of TD.

Keywords traveler's diarrhea; phage-based probiotic dietary supplement; phage prophylaxis; ETEC; EAEC; EHEC

1. Background and aims

Traveler's diarrhea (TD) is a polyetiological clinical syndrome characterized by three or more unformed stools a day in people traveling into another country or climate zone, particularly in tourists. All of TD cases are caused by microorganisms entering the human body with fecally-contaminated food and water. Dysfunction of gastro-intestinal tract in tourists most frequently appears within the first two weeks after arrival. During the trip, depending on the region, 25% to 75% of tourists experience one or more episodes of diarrhea. Usually, TD doesn't last longer than one week, however, in 6-10% of cases it may continue for two weeks or longer. TD may be caused by changes in food and regimen in a new place of stay. Around 100 million people from developed countries travel annually to regions with a high risk of developing TD: Latin America, South-East Asia and Africa, which results in up to 40 million cases of TD per year. More than 80% of traveler's diarrhea cases are caused by a variety of bacterial enteropathogens: diarrhea-producing E.coli accounts for the majority of infections (15-40%), although Campylobacter (3-20%), Salmonella (3-15%), Shigella (3-10%) and some other bacteria have also been isolated from travelers [1]. It is known that probiotics may offer a safe and effective method for TD prevention [2]. Abedon T.S. et al. in accordance with the definition of probiotics given by the World Health Organization, where they are described as having a healing effect on human organism when administered in adequate quantities, consider that it is possible to classify bacteriophage-based preparations as a new class of probiotics – phagebiotics [3].We have developed a phage-based probiotic dietary supplement for prophylaxis of TD caused by E.coli (including ETEC, EAEC and EHEC strains), Shigella flexneri, Shigella sonnei and Salmonella enterica. We have evaluated its effectiveness against infection caused by the non-pathogenic Lac (-) strain of E. coli in animal and human trials.

2. Materials and Methods

Characterization of the phage-based probiotic dietary supplement consisting of seven bacteriophage strains active against the world's most wide-spread TD pathogens was published by the authors earlier [4]. It reflected the origin of the phages, their lytic spectrum, maximum cultivation titers, molecular genetics analysis, including PCR-confirmation of absence of known pathogenic loci. Their virulent nature, proved by both the absence of known integrases and using the on-line program PHACTS (www.phantome.org/PHACTS) and their difference from earlier known relative strains of bacteriophages based on a full genome sequencing (Ion Torrent semiconductor sequencing) and bioinformatics analysis carried out in the Ubuntu 12.04 operating system based on the Linux core. De-novo automatic genome assembly was performed using NEWBLER software. The obtained data was visualized using the BRIG software [5]. DNA of coliphages: EcD7 and V18, contained in the phage-based probiotic dietary supplement, compared with already known representatives of Myoviridae families are presented in Fig.1-2.

Using the method of model infection suggested by Drozdova O.M. et al. [6], we conducted the following pilot experiment on 40 lab animals and 45 healthy volunteers: the control groups were infected with a non-pathogen strain of E. coli K12C600 during 3 days in doses of 5×10^7 CFU/ml every 24 hours, and the study groups were subjected to identical infection while taking the phage-based probiotic dietary supplement before, during and 24 hours after the infection period. The animals and healthy volunteers selected for the experiment had no signs of presence of lactose-negative strains of E. coli in their initial stool samples. The animals and humans in the control groups took saline solution instead of the phage-based probiotic dietary supplement following the same pattern (Table 1). To ensure a maximum accuracy of the experiment, the data of the microbiological study were verified by the molecular genetics method. E. coli cultures obtained from each dilution of the feces were washed off the surface of the dishes with a sterile saline solution for subsequent extraction of DNA from the bacterial suspension via sorption method and performing a PCR specific to the E. coli strain K12 C600 with end-point detection. We used the primers for the amplification which were previously described by Kuhnert P. et al. [7].

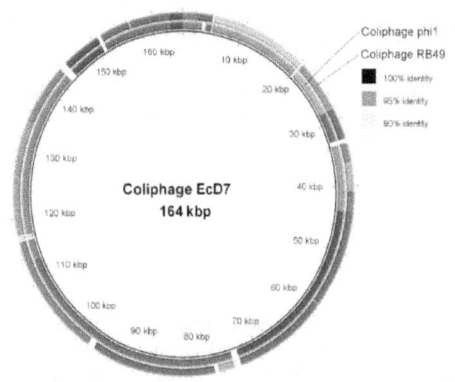

Fig. 1 A comparison of the genome of the coliphage EcD7 with the affine phages of the T4-like viruses Phi1 and RB49.

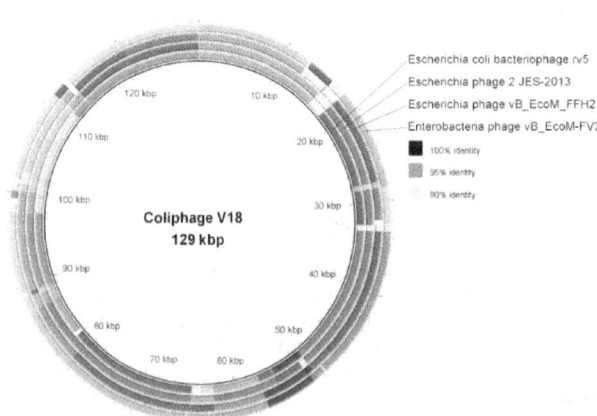

Fig. 2 A comparison of the genome of the coliphage V18 with the affine phages of the Myoviridae rv5, 2 JES-2013, vB_EcoM_FFH2 and vB_EcoM-FV3.

3. Results

The results of the trials are presented in the Table 2. The feces of the control groups showed lactose-negative E. coli on the 6th day of the experiment (Fig. 3a). At the same time, in the feces of the animals and volunteers from the study groups only E. coli with normal fermentation properties was found (Fig. 3b).

It is worth noticing, that in the stool samples obtained on the 6th day from the animals and humans of the study groups bacteriophages were present in 100% of cases. Seventy-two hours after the last intake of the phage-based probiotic dietary supplement, coliphages were found in only 20% of the samples.

Table 3 shows electrophoregrams, allowing to visualize the results of the amplification (horizontal electrophoresis of a PCR-product). Before the infection, the E. coli strain K12 C600 was absent in all of the stool samples. On the 6th day of the experiment E. coli K12 C600 was present in the persons of the control group up until the fifth dilution of the feces but was not found in the samples from the study group.

Table 1 Design of a pilot study on modeling of E.coli TD.

Groups	Study	Control
Lab animals – mice		
Number	20	20
Initial control for absence of lactose negative E. coli on the 1st day of experiment		
Infestation on 2nd-4th day of experiment	Once in 24 hours intragastrically 0.5 ml E. coli K12 C600 in titer 5×10^7 CFU/ml	
Prophylaxis from 1st to 5th day	Once in 24 hours 0.5 ml, phage-based probiotic dietary supplement	Once in 24 hours 0.5 ml, saline solution
Microbiological analysis - identification of lactose negative E. coli with subsequent PCR-analysis using specific primer for E.coli K12 C600 on 6th day		
Humans – healthy volunteers		
Number	30	15
Initial control for absence of lactose negative E. coli on the 1st day of experiment		
Infestation on 2nd-4th day of experiment	Once in 24 hours per os 20 ml E. coli K12 C600 in titer 5×10^7 CFU/ml	
Prophylaxis from 1st to 5th day	Twice in 24 hours 50 ml, phage-based probiotic dietary supplement	Twice in 24 hours 50 ml, saline solution
Microbiological analysis - identification of lactose negative E. coli with subsequent PCR-analysis using specific primer for E.coli K12 C600 on 6th day		

Table 2 Microbiological test results for feces.

Groups	1st day (prior to infestation)	6th day
Identified microorganisms	Lactose-negative E. coli, CFU/g	
Lab animals – mice		
Control	0	≤ 10^4
Study	0	0
Humans – healthy volunteers		
Control	0	≤ 10^6
Study	0	0

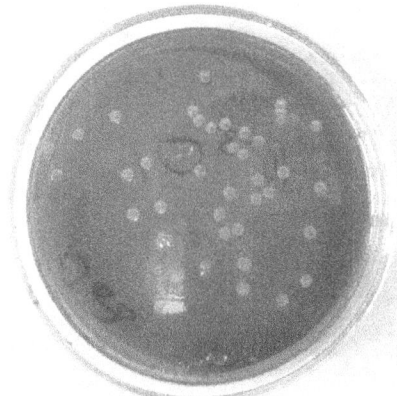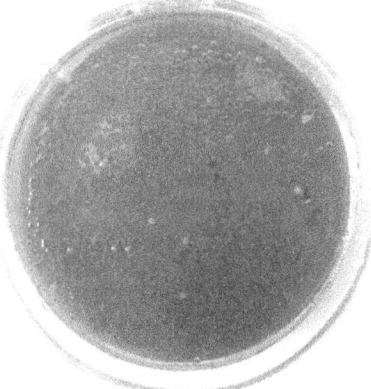

a)   b)

Fig. 3 Lactose-negative E. coli from the feces of the control a) and lactose-positive E. coli from the study b) groups on the 6th day of the experiment (Endo agar).

Table 3 PCR-test results.

Primers for amplification	K12-R (59-ATCCTGCGCACC AATCAACAA-39), K12IS-L (59-CGCGATGGAAGATGCTCTGTA-39). Amplicon length 969 bp [7].	
Day of experiment	1 (prior to infestation)	6
Study groups	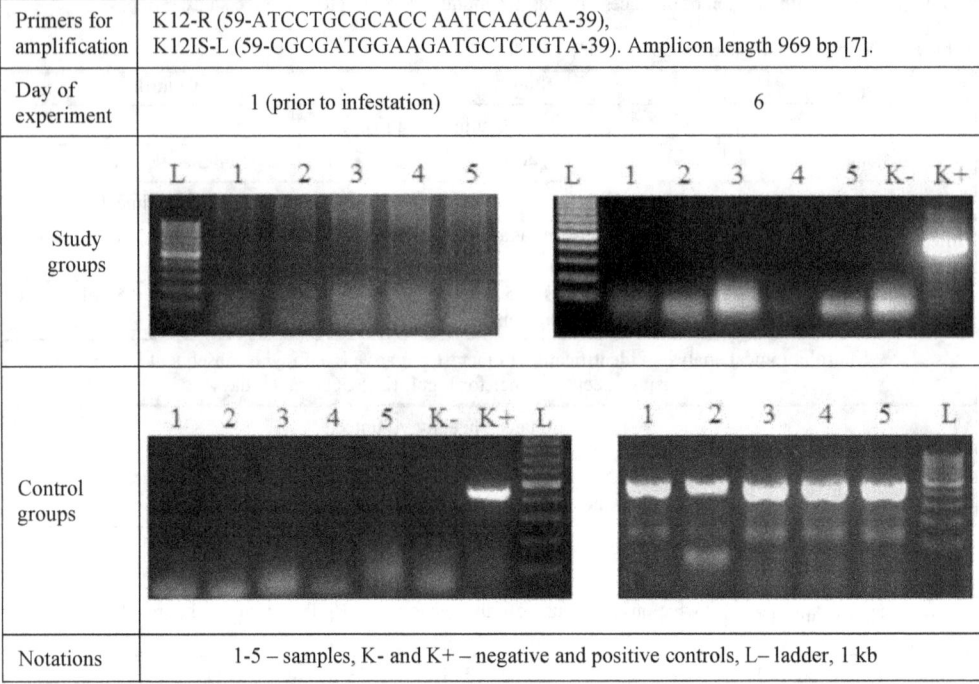	
Control groups		
Notations	1-5 – samples, K- and K+ – negative and positive controls, L– ladder, 1 kb	

4. Conclusions

Thus, the experimental model of an infectious process on lab animals and a small group of volunteers proves possible specific prophylaxis of traveler's diarrhea caused by E. coli, using the phage-based probiotic dietary supplement. These results have great significance because two original coliphages included in this phage-based probiotic dietary supplement have a broad host-range including O26, O55, O103, O104, O121, O125, O127, O128, O145, O146, O157 E. coli strains which cause severe cases of TD.

References

[1] Diemert DJ. Prevention and self-treatment of traveler's diarrhea. Clinical Microbiology Reviews. 2006; 19(3):583-594.
[2] McFarland LV. Meta-analysis of probiotics for the prevention of traveler's diarrhea. Travel Medicine and Infectious Disease. 2007; 5:97-105.
[3] Abedon ST, Kuhl SJ, Blasdel BG, Kutter EM. Phage treatment of human infections. Bacteriophage 2011; 1(2):66-85.
[4] Aleshkin AV, Volozhantsev NV, Svetoch EA, Verevkin VV, Krasilnikova VM, Bannov VA, et al. Bacteriophage cocktail in prophylaxis against food-borne infections. In: Mendez-Vilas A, editor. Worldwide research efforts in the fighting against microbial pathogens: from basic research to technological developments. Boca Raton: BrownWalker Press; 2013. p. 100-104.
[5] Alikhan N-F, Petty NK, Ben Zakour NL, Beatson SA. BLAST Ring Image Generator (BRIG): simple prokaryote genome comparisons. BMC Genomics. 2011; 12:402.
[6] Drozdova OM, An RN, Chanishvili TG, Livshits ML. Experimental study of the interaction of phages and bacteria in the environment. Journal of Microbiology, Epidemiology and Immunobiology. 1988; 7:35-39.
[7] Kuhnert P, Nicolet J, Frey J. Rapid and accurate identification of Escherichia coli K-12 strains. Applied and Environmental Microbiology. 1995; 61(11):4135-4139.

Formulated natural plant extracts from nutmeg and cardamom show antifungal activity against clinical isolates of *Candida albicans* and affects cellular morphology and ergosterol

V. S. Radhakrishnan[1], Adeppa Kuruba[2], Surya Prakash Dwivedi[3] and Tulika Prasad[1,*]

[1]Advanced Instrumentation Research Facility (AIRF), Jawaharlal Nehru University, New Delhi-110067, India
[2]India Pesticides Limited, Lucknow, U.P., India
[3]School of Biotechnology, IFTM University, U.P, India
[*]Corresponding author email: prasadtulika@hotmail.com; prasadtulika@mail.jnu.ac.in

Candida is an opportunistic fungal pathogen accounting for high rates of mortality and morbidity in immunocompromised individuals. Of all *Candida* spp, *Candida albicans* is the fourth most common cause of nosocomial infections globally. Emergence of multidrug resistance and limitations in availability of broad spectrum drugs with minimum host side effects has been hindering the control and treatment of fungal infections. In the recent times, natural products have gained impetus in their use as antimicrobials owing to their broad spectrum, multiple targets in the microbial cells and negligible host toxicity. For this study we prepared formulations of two plant products, nutmeg and cardamom. The formulations were prepared by extracting the active components using isopropanol followed by subsequent distillation and emulsification by ethylacetate. Anti-*Candidal* activity of the formulations of these two extracts were confirmed by both broth microdilution assay and spot assays. The growth inhibition was found to be more in nutmeg than cardamom. Formulations of nutmeg and cardamom revealed more than 80% growth inhibition at respective concentrations of 50µl/ml and 100µl/ml (v/v). Treated cells showed reduced ergosterol content, altered cell morphology, cell wall thickening, membrane aberrations and cytoplasm displacement. The antimicrobial effects observed may be due to the phenols, polyphenols, terpenoids, flavonoids, alkaloids and quinones present in the plant extracts. Further investigations are underway. This study has the potential to contribute to the development of new therapeutic strategies for clinical applications in the treatment of *Candidiasis*.

Keywords: *Candida albicans;* antifungal agents; natural plant products; ergosterol; cardamom; nutmeg

1. Introduction

In the estimated 8.7 million eukaryotic species of universe, 600 different fungal species are considered to be pathogenic to human and cause various infections ranging from common, superficial to life threatening invasive infections [1]. The mortality rate for people suffering from top 10 fungal diseases is equal to or higher than those suffering from TB or malaria [2]. *Candida* is among the top four fungal genera which together result in more than 90% of all reported fungal-related deaths. *Candida albicans* dominates in incidence among all *Candida* spp and is the fourth most common cause of health-care related blood stream infections [3, 4]. *Candida albicans* otherwise is a commensal, dimorphic, opportunistic pathogen which is present in various parts of the human body such as gastrointestinal, urinary tracts etc as part of normal human microbiome [4].

Several classes of antifungal drugs are available for the treatment of fungal infections by targeting the fungal cell wall (echinocandins - caspofungin), ergosterol (polyenes - amphotericin B and nystatin), ergosterol biosynthesis (azoles – fluconazole, itraconazole, ketoconazole, voriconazole, posoconazole and allylamines - terbinafine) and nucleic acid biosynthesis (pyrimidine analogs - flucytosine) [2, 5, 6].

Initial lag due to delayed disease diagnosis, fungal identification and further restrictions in route of administration, spectrum of activity and bioavailability in target tissues serve as a major roadblock in successive treatment of fungal diseases [2]. Antifungal therapy has been further complicated by detrimental drug interactions, host toxicity and emergence of drug resistance. With the increase in use of antimicrobial agents, the prevalence of antimicrobial drug resistance has evolved as an alarming threat to global public health [6]. Although bacterial drug resistance is the primary cause of concern but rise in antifungal resistance has also begun to pose a challenge to the therapeutics [4].

Multidrug resistance (MDR) is a multifactorial phenomenon and various mechanisms (Fig. 1) such as molecular alteration of drug target, drug inactivation, up-regulation of target enzyme/ gene amplification, overexpression of drug efflux pumps and altered drug accumulation inside the cells by either changes in the membrane permeability constraints or altered facilitated drug diffusion [7, 8] have been reported to contribute to the development of MDR.

There are relatively fewer classes of antifungal drugs available which is further restricted by emergence of MDR. Clinically significant drug resistance in *Candida albicans* has been reported in immunocompromised individuals especially those undergoing long term antifungal therapy [5]. The mechanisms resulting in high-level clinical resistance may be suppressed either by inhibiting the drug efflux pumps or affecting other cellular targets to chemosensitize the cells to already existing drugs [6]. This paves the way for the need to find novel drug targets and new cheaper inhibitors with low host toxicity and broad spectrum activity for effective antifungal therapy in the future.

This study aims to determine the anti-*Candidal* activity of formulated plant extracts of nutmeg and cardamom to decipher their mode of action for their future scope of use either singly or in combinatorial therapy with other drugs with an ultimate aim for better and improved therapeutic strategy for MDR reversal.

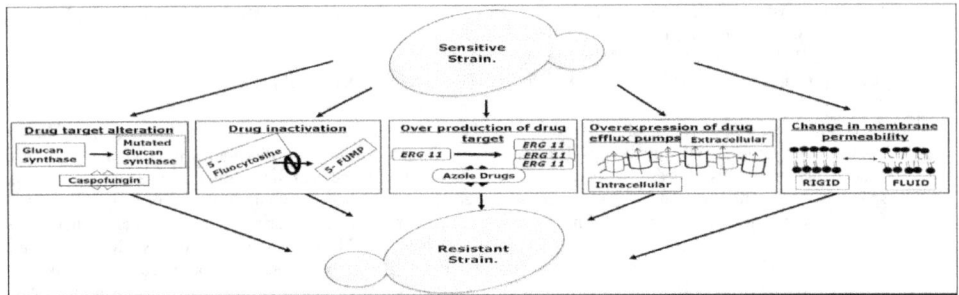

Fig. 1: Schematic representation of drug resistance mechanisms.

2. Materials and methods

2.1 Materials, strains and growth media:

Media chemicals were obtained from HI-Media (Mumbai, India) and Fisher Scientific (Mumbai, India). Analytical grade EDTA, Sorbitol, Tris-Cl and MgSO$_4$ were obtained from Qualigens Chemicals, India. All solvents including *n*-Heptane, Methanol, Ethanol, Ethyl Acetate, Isopropanol, water were of HPLC grade. Enzyme Lyticase, Pyrogallol, (DMSO), Tetrahydrofuran (THF), ergosterol standard, fluorescent probes 1, 6-Diphenyl 1, 3, 5-hexatriene (DPH) were procured from Sigma Chemical Co. (St. Louis, Mo., USA).

CAF2-1 [9] has been used for this study and maintained on liquid broth containing 2% (w/v) Dextrose, 1% (w/v) Yeast extract and 2% (w/v) Peptone (YEPD medium) and 2.5% (w/v) Agar plates for solid media. Cells were grown at 30°C with continuous shaking at 140-150 rpm for 14-16 hrs and these exponentially growing cells were used for all the experiments.

2.2 Extract preparation and analysis of the phyto-constituents:

Cardamom and nutmeg collected from Southern India were ground into fine powder and active components of plant extracts were first extracted by isopropanol and then concentrated by distillation method. Finally the concentrated extract was emulsified using ethyl acetate. Extract was subjected to detailed GC-MS analysis for identifying the active components present in the extract. Metabolite profiling of plant extracts was performed on Trace GC ultra (Thermo Scientific, FL, USA) coupled to TSQ Quantum XLS mass spectrometer (Thermo Scientific, FL, USA). TG-5MS fused silica capillary column (30 m x 250 μm i.d.; Thermo Scientific), chemically bonded with 5% phenyl 95% methyl polysiloxane cross linked stationary phase (0.25μm film thickness) was utilized to separate the peaks. Chemical components from the extracts were identified by comparing with authentic compounds using WILEY8.LIB and NIST05.LIB.

2.3 Drug susceptibility tests:

Susceptibility to the plant extracts were estimated by both broth microdilution and spot assay method. Growth control was maintained in absence of plant extracts. MIC was determined by the broth microdilution method as described earlier in accordance with the recommendations of the Clinical and Laboratory Standard Institute (CLSI), formerly National Committee for Clinical Laboratory Standards (NCCLS) [10, 11]. After incubation at 30°C for 48 hrs, the MIC end point for growth inhibition was determined by comparing with the growth of extract-free control. Spot assay was performed as described earlier [10, 12]. Growth differences were recorded after incubation of the plates for 48 hrs at 30 °C.

2.4 Assessing the physical state of the membrane:

Fluorescence polarization which gives a measure of membrane fluidity was assessed using fluorescent probe 1, 6-diphenyl- 1, 3, 5-hexatriene (DPH). The method earlier described [10, 13] was followed with little modifications. Spheroplasts were prepared by incubating the cells with lyticase at 30°C for 3-4 hrs. The spheroplasts were re-suspended in labelling buffer pH 7.4 (0.6M Sorbitol, 10mM MgSO$_4$ and 20mM Tris-Cl) and incubated with 2μm DPH for 1hr at 30°C. The polarization ratio (p) is unit less and calculated on Perkin-Elmer LS55 spectrofluorimeter (excitation 360 nm, emission 426 nm and slit size 10 nm for both excitation and emission) as follows [14]:

$$p = \frac{I_{VV} - (I_{VH} \times G)}{I_{VV} + (I_{VH} \times G)}$$

where, I_{VV} = Corrected fluorescence intensity obtained with excitation by vertically polarized light and emission detected by analyzer oriented vertically to the direction of polarized excitation light, I_{VH} = Corrected fluorescence intensity obtained with excitation by vertically polarized light and emission detected by the analyzer oriented horizontal to the direction of polarized excitation light. *Grating Factor G* , the correction for optical components of the instrument is calculated as I_{HV} / I_{HH} where subscripts HV and HH indicate the corrected fluorescence intensity values obtained with horizontal-vertical and horizontal-horizontal orientations for the polarizer and analyzer in that order respectively.

2.5 Sterol estimation:

Sterols were extracted as described earlier [13, 15] with slight modifications from cells grown overnight at 30°C in the absence (control) and presence of extracts. Cell pellets were re-suspended in 2.5 ml methanol, 1.5 ml potassium hydroxide (60% [wt/vol]) and 1 ml methanol dissolved in pyrogallol (0.5% [wt/vol]). The suspension was kept for refluxing at 80°C for 2 hrs and allowed to cool. The sterol was extracted with the help of *n*-heptane and extraction was repeated by adding known volume of *n*-heptane two to three times. The extracted sterols indicated four-peak spectral absorption patterns which are produced by ergosterol and 24(28)-dehydroergosterol (24(28)-DHE) contents. Both ergosterol and 24(28)-DHE absorb at 281.5 nm, whereas only 24(28)-DHE absorbs at 230 nm. Ergosterol content was determined by subtracting the amount of 24(28)-DHE (calculated from the A_{230}) from the total ergosterol + 24(28)-DHE content (calculated from the $A_{281.5}$).

2.6 Study of surface morphology and ultrastructure of *Candida* cells by electron microscopy:

Scanning electron microscopy (SEM) and Transmission electron microscopy (TEM) were used to monitor the ultrastructural and surface morphological alterations respectively [16] after treatment with extracts. Cells grown with and without (control) sub-inhibitory concentration of each extract (4μl/ml) were harvested, washed with PBS buffer and fixed with 2% glutaraldehyde in 0.1M phosphate buffer for 3 hrs at 25°C. After washing with phosphate buffer (pH 7.2), were post-fixed with 1% OsO$_4$ in 0.1 M phosphate buffer for 1 hr at 4°C for further study under SEM and TEM. For SEM, an aliquot of the above treated cells were dehydrated in acetone and dropped on round glass cover slip with hexamethyldisilizane (HMDS) and dried, followed by sputter coating with gold for observation under SEM (Zeiss EVO40 with Resolution: 3.0 nm @ 30 kV (SE and W) and Magnification 7 to 1,000,000x). For TEM, samples were dehydrated with graded acetone, cleared with toluene and infiltrated with toluene and araldite mixture at 25°C and finally put overnight in pure araldite at 50°C. The samples are then embedded in 1.5 ml Eppendorf tube with pure araldite mixture at 60°C for semi thin and ultrathin section cutting with ultra-microtome (Ultra-microtome Leica EM UC6). Sections taken on 3.05 mm diameter and 200 mesh copper grid were stained with uranyl acetate and observed at 120 kV under TEM (Model name JEOL2100F with Accelerating Voltage 80kV-200kV and Magnification 50 to 1,500,000 X).

3. Results and Discussion

Invasive fungal infections, growing spectrum of fungal pathogens, limited availability of therapeutic options and development of MDR together constitute a significant burden in patients with impaired immunity [7]. Among various fungal pathogens, *Candida* spp have gained remarkable significance due to their increasing prevalence in various patient groups [17]. Although non-*albicans Candida* have been identified among four most common blood stream pathogens, *Candida albicans* is still the most frequently isolated species from tissue and blood samples [17].

Clinical needs for novel antifungal agents with broad spectrum activity and minimal host toxicity has encouraged the search for novel therapeutic alternatives [18]. Plant products are known to be rich sources of bioactive molecules with antimicrobial properties [18]. Owing to the long history of natural plant products in traditional medicine and their safe human use, recent times have seen the increasing popularity of trying to develop alternative therapeutic molecules from plant products [18]. Such studies still have scant credibility and limited knowledge exists till date related to antimicrobial activity and the mechanism of action of plant products. Therefore, methodical and detailed investigation is required to address such issues to translate the application of a natural plant product into clinical therapeutics.

This study is an effort to characterize the mode of antifungal action of formulated extracts of nutmeg and cardamom against opportunistic fungal pathogen, *Candida albicans*. The composition of the phyto-constituents was determined by gas chromatography and mass spectrometry. A total of 29 (viz. sabinene, isoeugenol, eugenol, borneol, cymene, myrcene, terpinene, terpineol etc.) and 14 (viz. cineole, α-terpineol, sabinene, ß-phellandrene etc.) components were identified to be present in nutmeg and cardamom respectively. Taking into account the known activities of some of the constituents present in nutmeg and cardamom or similar constituents in other plant sources, it was pertinent to determine the respective phyto-constituents.

Broth micro-dilution assay and spot assay were used to elucidate the anti-*Candidal* activity of both plant extracts. Both nutmeg and cardamom were found to inhibit growth of *Candida* cells in a dose dependent manner. Broth microdilution assay (Fig. 2A) revealed the Minimum Inhibitory Concentration (MIC_{80}) for nutmeg (■) and cardamom (●) for *Candida* cells at 25µl/ml and 100µl/ml (v/v) respectively on comparison with growth control (▲) (absence of extracts). Spot assay (Fig. 2B) for plant extracts at the concentration of 40µl/ml confirmed the fungicidal activity of both plant extracts. The extracts formulated were emulsified in ethyl acetate and no inhibitory effect of ethyl acetate alone was observed.

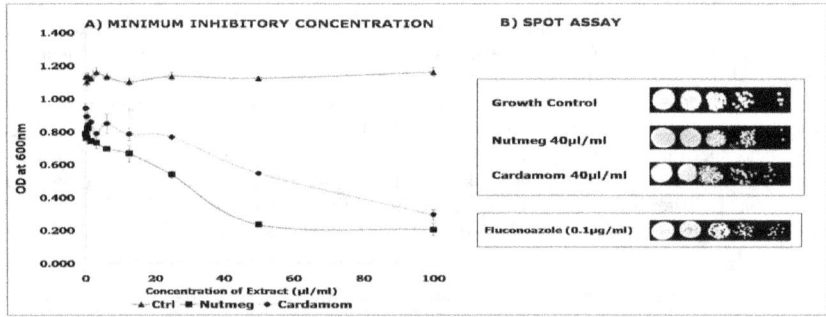

Fig. 2: Fungicidal activity of formulations of nutmeg and cardamom as determined by broth microdilution assay and spot assay: **(A)** MIC_{80} for *Candida* cells observed at 25µl/ml and 100µl/ml for nutmeg and cardamom respectively; **(B)** Spot assay.

Ergosterol content of the *Candida* cells were found to be reduced after treatment with the formulated extracts (Fig. 3A). Sterols are important structural and regulatory components of eukaryotic cell membranes. Ergosterol, the final product of sterol biosynthetic pathway in yeast is responsible for cellular functions and structural features such as signal transduction, permeability and fluidity [19]. Treatment with nutmeg and cardamom formulations resulted in 25% and 5% reduction of cellular ergosterol levels respectively. Observed altered membrane order/ physical state of the membrane (Fig. 3B) may be attributed to the reduction in the sterol content and this was reflected by the reduced fluorescence polarization 'p' values for the treated cells. Increase in membrane fluidity by 18% and 8% in the cells was observed in cells treated with formulations of nutmeg and cardamom respectively.

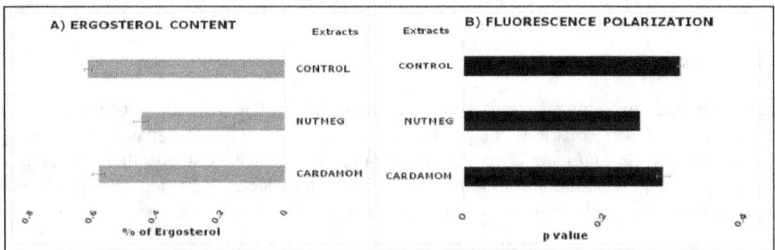

Fig. 3: (A) % ergosterol content ± the standard deviation of the mean of the three sets of experiments; **(B)** Mean fluorescence polarization "p" values of the cells (inversely proportional to membrane fluidity) ± the standard deviation of the mean of the three sets of experiments; in the control (untreated) and treated cells.

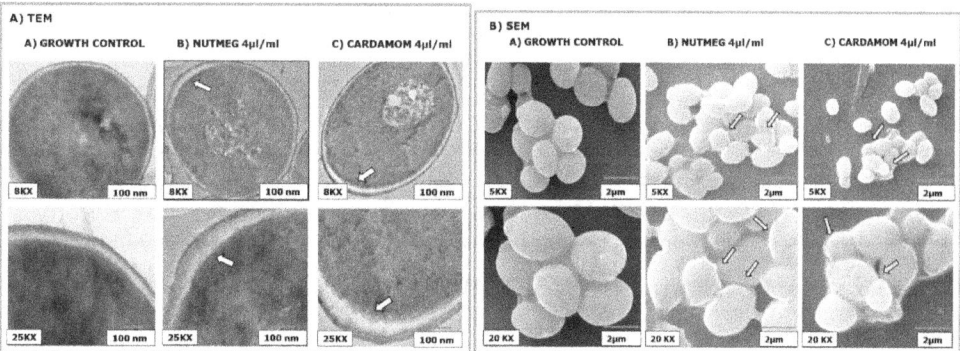

Fig. 4: (A) TEM images (taken at 8KX and 25KX magnification); **(B)** SEM micrographs (taken at 5KX and 20KX magnification); for untreated (growth control) and cells treated with formulations of nutmeg and cardamom.

Treatment with even sub-inhibitory concentration (4µl/ml) of extracts of nutmeg and cardamom resulted in morphological changes and alteration of cellular ultrastructure as observed by electron microscopy (Fig. 4A and B). Ultrathin sections of treated cells (Fig. 4A) revealed increased cell wall thickness and cytoplasmic displacement. SEM (Fig. 4B) showed treated cells with corrugated and shrunken cell surfaces as compared to smooth surfaces of untreated cells.

It is evident from the results that formulations prepared from nutmeg and cardamom have multiple cellular targets and appear to have the potential to contribute to development of new therapeutic strategies for clinical applications. Further investigations are underway. This study can help to correlate this data for other pathogenic fungi of medical importance. Plant derived products have gained importance to use as antimicrobial agents because of their multi targeted action and low or negligible host toxicity. Combinations of drugs may prove to be more efficacious than single drug therapy, and therefore newer and cheaper drugs with broad spectrum activity and minimal host toxicity are needed to overcome the challenges of fungal therapeutics [6].

Acknowledgements: This work has been supported by grants to TP from Department of Biotechnology (BT/PR5110/MED/29/497/2012 and BT/BI/12/045/2008), University Grants Commission UPE II scheme and Department of Science and Technology (DST-PURSE), India. JNU and AIRF for providing infrastructural support and Department of Biotechnology (DBT), India for the award of Senior Research Fellowship to VSR are gratefully acknowledged.

References

[1] Mayer FL, Wilson D and Hube B. *Candida albicans* pathogenicity mechanisms. Virulence. 2013; 4(2):119–128.

[2] Brown GD, Denning DW, Gow NAR, Levitz SM, Netea MG and White TC. Hidden Killers: Human Fungal Infections. Sci Transl Med. 2012; 4(165):1-9.

[3] Morgan J. Global Trends in *Candidemia*: Review of Reports from 1995–2005. Curr Infect Dis Rep. 2005; 7:429–439.

[4] Wisplinghoff H, Bischoff T, Tallent SM, Seifert H, Wenzel RP and Edmond MB. Nosocomial Bloodstream Infections in US Hospitals: Analysis of 24,179 Cases from a Prospective Nationwide Surveillance Study. Clinical Infectious Diseases. 2004; 39:309–17.

[5] Odds FC, Brown AJP and Gow NAR. Antifungal agents: mechanisms of action. Trends Microbiol. 2003; 11(6):272-279.

[6] Cannon RD, Lamping E, Holmes AR, Niimi K, Tanabe K, Niimi M and Monk BC. *Candida albicans* drug resistance – another way to cope with stress. Microbiology. 2007; 153:3211–3217.

[7] Kanafani ZA and Perfect JR. Resistance to Antifungal Agents: Mechanisms and Clinical Impacts. Clinical Infectious Diseases. 2008; 46:120–128.

[8] Prasad T, Sethumadhavan S and Fatima Z. Altered Ergosterol biosynthetic pathway - an alternate multidrug resistance mechanism independent of drug efflux pump in human pathogenic fungi *C. albicans*. Science against microbial pathogens: communicating current research and technological advances. Formatex Microbiology series. 2011; 3:757-768.

[9] Fonzi, WA, and Irwin MY. Isogenic strain construction and gene mapping in *Candida albicans*. Genetics. 1993; 134:717–728.

[10] Prasad T, Chandra A, Mukhopadhyay CK, and Prasad R. Unexpected Link between Iron and Drug Resistance of Candida spp.: Iron Depletion Enhances Membrane Fluidity and Drug Diffusion, Leading to Drug-Susceptible Cells. Antimicrob. Agents Chemother. 2006; 50(11):3597–3606.

[11] Wayne, PA. Reference Method for Broth Dilution Antifungal Susceptibility Testing of Yeasts. Approved Standard M27-A3, third ed. Clinical and Laboratory Standards Institute. CLSI. 2008.

[12] Mukhopadhyay K, Prasad T and Prasad R. Membrane sphingolipid and ergosterol interactions are important determinants of multi drug resistance in *Candida albicans*. Antimicrob. Agents Chemother. 2004; 48(5):3695–3705.

[13] Prasad T, Hameed S, Manoharlal R, Biswas S, Mukhopadhyay CK, Goswami SK and Prasad R. Morphogenic regulator *EFG1* affects the drug susceptibilities of pathogenic *Candida albicans*. FEMS Yeast Res. 2010; 10:587–596.

[14] Shinitzky M and Barenholz Y. Fluidity parameters determined by fluorescence polarization. Biochim. Biophys. Acta. 1978; 515:367–394.

[15] Martel CM, Parker JE, Bader O, Weig M, Gross U, Warrilow AGS, Kelly DE and Kelly SL. A Clinical Isolate of Candida albicans with Mutations in *ERG11* (Encoding Sterol 14α -Demethylase) and *ERG5* (Encoding C22 Desaturase) Is Cross Resistant to Azoles and Amphotericin B. Antimicrob. Agents Chemother. 2010; 54(9):3578.

[16] Walker LA., Munro CA., Bruijn Id, Lenardon MD, McKinnon A and Gow NAR. Stimulation of Chitin Synthesis Rescues *Candida albicans* from Echinocandins. PLoS Pathogens. 2008; 4(4): e1000040:1-12.

[17] Weig M and Brown AJP. Genomics and the development of new diagnostics and anti-*Candida* drugs. Trends Microbiol. 2007; 15(7):310-317.

[18] Cowan MM. Plant products as antimicrobial agents. Clin. Microbiol. Rev. 1999; 12(4):564-582.

[19] Veen M and Lang C. Interactions of the ergosterol biosynthetic pathway with other lipid pathways. Biochem. Soc. Trans. 2005; 33(5):1178 – 1181.

Investigation of the antituberculous effect in vivo of the new remedies

V.L. Bismilda[1], L.T. Chingisova[1], T. S. Zveryachenko[2], A. S. Zveryachenko[3]

1. National Center for Tuberculosis Problems, Almaty, Republic of Kazakhstan
2. North-Kazakhstan State University named after M. Kozybaev, Petropavlovsk, Republic of Kazakhstan
3. Federal budget institutions "662 Center for medical equipment and property", Mytischi, Moscow region, The Russian Federation
*Corresponding author: e-mail: timurzv@mail.ru, Phone: +7 777 296 33 11

Abstract. The antituberculous activity of new remedy was studied by the experiments on guinea pigs. The best results were achieved in the group with a combination of traditional antibiotics and syrup. Mass recovery of the remaining animals was faster, and compared with the group treated with only traditional antibiotics, the weight of animals of this group was about 20 g more by the end of the experiment. Also, as a result of treatment cough attacks were completely absent. 2 animals were completely healthy by the end of treatment, and the remaining 3 had minor lesions with minimal growth (+1) only in the inguinal lymph node closest to the site of infection. In the group by self-use of the syrop mass reduction of animals significantly slowed down and was $90 \pm 5,2$ g by the end of the experiment. Although physical activity was reduced, and there was a depression at the same time the amount of begma and cough attacks decreased significantly in comparison with the animals of the non-treated group and the culture growth of internal organs was 2+. Thus, we can conclude that the self-use (without antibiotics) in the studied dosage of the researched medication has bacteriostatic action against Mycobacterium tuberculosis in the experiment in vivo, thus effectively reducing the amount of begma and cough attacks, which helps to reduce bacterioexcretion.

Keywords: aspen, wartwort, licorice, tuberculosis.

Introduction

The new drug (syrup) was developed. It is based on extracts of wartwort, aspen bark and licorice root, aimed for oral use in the treatment of inflammatory and infectious diseases of the respiratory organs, complex therapy of pulmonary tuberculosis and treatment of accompany to tuberculosis pathogenic flora.

The main components of syrup are wartwort alkaloids, phenolic compounds of aspen, an extract of licorice. Syrop was investigated in vitro concerning strains of tuberculosis ($H_{37}Rv$ and 2 antibiotic-resistant strains), and also strains of Staphylococcus aureus and Streptococcus Pneumonia. The investigated cultures were highly resistant to traditionally used antibiotics. The accurate cidal effect was established. There was also studied antitussic activity and safety of preparations in acute and long-term experiment [1].

Object of the research

The new dosage form (syrup) based on extracts of wartwort (Chelidonium majus), aspen bark (Populus tremula L.) and licorice root (Glycyrrhiza uralensis Fisch), aimed for oral use in the treatment of inflammatory and infectious diseases of the respiratory organs and complex therapy of pulmonary tuberculosis.

The aims of the work

The research of the antituberculous activity (strain H37Rv) by the experiments on guinea pigs.

Materials and methods

The experiment was conducted on 16 males and 16 females guinea pigs (weighing 250-300 g). Animals were challenged with a museum strain of Mycobacterium tuberculosis $N_{37}R_v$.

The research was conducted at the National Reference Laboratory of the National Center of Tuberculosis problems (Almaty, Kazakhstan) in accordance with the guidance [2,3,4].

The experimental study was performed on 32 guinea pigs with the weight 250-300 g, which were kept in the same conditions and the same type of diet throughout the experiment.

Animals were divided into 5 groups (the 2nd Group included 8 animals, the others included 6):
  1. Control uninfected animals.

2. Control infected animals without treatment.

3. Control infected animals, treated with conventional antibiotics (rifampicin – 10,0 mg / kg, isoniazid – 15,0 mg / kg per orally).

4. Experience infected animals treated with antibiotics (rifampicin - 10.0 mg / kg of isoniazid - 15.0 mg / kg, per orally) and with the study drug.

5. Experience infected animals, with the study drug treatment.

Animals of groups 4 and 5 treated 0,125 ml of the studied drug (that is 0,5 ml / kg of body weight) diluted with water 1:10 (i.e., 0,125 ml of the drug + 1,25 ml water) per orally 3 times a day.

According to this method of the dose conversion the dosage for guinea pigs corresponds to the dosage for an adult (weighing 80 kg) 0,06 ml / kg.

Thus, a dose of 0,125 ml for guinea pigs weighing 0,25 kg according to the method is equivalent to 5 ml for an adult person weighing 80 kg.

Animals of groups 2-5 were challenged by subcutaneous injection into the right inguinal part of 0,1 mg in 0,5 ml of saline per individual with strain $H_{37}R_v$. At this dose of infection the death of 95-100% of the animals takes place by the end of the 1st month after infection.

Treatment of animals was started after the development of tuberculous process, which was tested by the general state (decreased activity, difficulty in breathing, weight loss) and by the results of histological, pathological and microbiological studies. The same parameters were monitored in experimental animals. Treatment was started on the 20th day after infection.

All the data were subjected to statistical analysis by standard methods.

Results

Animals of the first group (uninfected) during the whole experiment retained their normal motorial and behavioral activity. Feed intake corresponded to normal. During the experiment the animals weight increased by 90 ± 6 g.

Animals of the second group which received no treatment died within 36 days after infection. In the experiment, they had the full range of behavioral, morphological and microbiological pathological changes attributable to tuberculosis.

Animals in this group suffered from infection most severely. In this group during the experiment there was noted 100% mortality of animals. All animals were sluggish, their fur was dull, by the end of the observation steadily increasing exhaustion and weakness led to the fact that guinea pigs were constantly lying, got up with difficulty, coughing bouts went into chronic with flux of phlegm. Mass loss during the experiment was about $100 \pm 3,1$ g.

It was observed the formation of a defect of the skin in animals of this group to the site of infection. At autopsy of dead animals it was observed plethora of internal organs, on the surface of the lungs, hepatic and spleen; numerous grayish-yellow mons were observed in many places merging with each other. Right inguinal lymph nodes were significantly increased in size, on the discission there was creamy pus. Inoculation of the animals in this group showed a heavy growth of Mycobacterium tuberculosis (3 +) from all samples.

Thus, the animals of the second group which did not receive the treatment experienced a tuberculous process with involvement of almost all internal organs, confirmed by macroscopic and microbiological studies.

In the third group (was treated with rifampicin – 10,0 mg/kg, isoniazid – 15,0 mg/kg, per orally) during the experiment only one female died on the 30th day after treatment. Mass reduction of the survived animals was $70 \pm 3,5$ g. At the autopsy3 animals had no pathological changes in the internal organs. Homogenate culture of internal organs of animals of this group gave a single growth of Mycobacterium tuberculosis (single colony from 1 to 5). Two animals had isolated small hillocks of grayish-yellow color, with no tendency to merge on the surface of the lungs and liver. Organs culture of these animals showed a slight increase (1+) of all samples.

Thus, these data confirmed the high protective effect of the essential anti-TB drugs (isoniazid, rifampicin) in the treatment of tuberculosis.

The best results were achieved in the fourth group with a combination of rifampicin (10,0 mg/kg) and isoniazid (15,0 mg/kg) with the test remedy (0,5 ml/kg) per orally. One animal in this group died on the 23rd day of treatment.

All the animals were active; they did not refuse the meals and hair loss was not observed. Since the beginning of the treatment it was observed the weight recovery by the animals, and by the end of the experiment the weight loss was only ($50 \pm 2,5$ g). Skin defect in the area of infection was not found. Coughing attacks during the treatment stopped.

At autopsy, two of the surviving animals had no pathological changes in the internal organs, the dead animal had increased inguinal lymph node to the site of infection, 2 times larger. On the discission there was yellowish liquid, characteristic for caseous necrosis. Lungs and liver were full-blooded, and their surface was covered with a small amount of little tubercles grayish color mons with no tendency to merge. Spleen and kidneys didn't have visible lesions. Remaining three animals also had an increase in the inguinal lymph node.

High cure rates in this group were confirmed by microbiological studies: 2 animals had no growth of Mycobacterium tuberculosis, 3 animals with enlarged lymph nodes had growth of Mycobacterium only 1 +.

In the fifth group by self-use of the researched remedy in a dosage of 0,5 ml/kg of body weight as well as in the third and fourth groups, only one animal died on the 23rd day of treatment, and the other animals survived till the end of the experiment (60 days). Mass reduction of animals significantly slowed down and was $90 \pm 5,2$ g by the end of the experiment. Although physical activity was reduced, and there was a depression at the same time the amount of begma and cough attacks decreased significantly in comparison with the animals of the 2nd group (non-treated).

All the animals were identified with all the macroscopic and microscopic signs of disease. At autopsy, all the animals of the fifth group had a plethora of internal organs, with the observed colonization of internal organs with tubercles gray-yellow mons, sometimes merging with each other.

The culture growth of internal organs was 2+.

Conclusion

Thus, we can conclude that the self-use (without antibiotics) in the studied dosage of the researched medication has bacteriostatic action against Mycobacterium tuberculosis in the experiment in vivo.

The rate of development of tuberculous process has fallen by more than 2 times compared with animals receiving no treatment.

It is advisable to explore higher doses of syrup for self use.

In the complex therapy with antibiotics the syrop effectively reduces the amount of begma and cough attacks, which helps to reduce bacterioexcretion.

References

[1] A. S. Zveryachenko, T. S. Zveryachenko The research of the safety of New Remedy for Complex Lung Tuberculosis Treatment in Acute and Long-term Experiment //Abstracts book of congress Phytopharm-2014. Saint-Petersburg: 2014, p. 67.

[2] Practical policies for the Study of antituberculous activity of pharmacological substances. Moscow: 2000. – 7 p.

[3] Руководство по экспериментальному (доклиническому) изучению новых фармакологических веществ. Под общей редакцией Р.У. Хабриева. Moscow: 2005. – 832 p.

[4] Руководством по лабораторным животным и альтернативным моделям в биомедицинских технологиях. Под редакцией Н.Н. Каркищенко, С.В. Грачева. Moscow: 2010. – 344 p.

Management of Chronic Periodontitis using Metronidazole local drug delivery device as an adjunct to subgingival debridement: A clinical, microbiological & molecular study

Lt Col Sangeeta Singh[1], Brig K K Lahiri[2], Brig Subrata Roy[3], Col A K Shreehari[4]

[1]MDS (Periodontology),Asst Prof, Dept of Dental Surgery, AFMC, Pune-40
[2]MD (Microbiology), Professor, Dept of Microbiology, AFMC, Pune-40,
[3]MDS(Periodontology),Professor,Command Military Dental Centre, Pune-40,
[4]MDS(Periodontology),Assoc Prof, Dept of Dental Surgery, AFMC, Pune-40

Background: Current trends in the management of inflammatory periodontal disease aim at altering the sub gingival ecosystems residing in the periodontal pockets in order to alter the pathogenic flora into a micro biota compatible with periodontal health. Systemic antimicrobial therapy is effective in controlling the infection, however it involves a relatively high dose with frequent doses over a prolonged period of time to achieve the required inhibitory concentrations in the sulcular fluid. The adjunctive use of local drug delivery may provide a beneficial response especially in specific areas where conventional forms of therapy might fail. The aim of this study was to evaluate the efficacy of a local drug delivery system containing Metronidazole as adjunct to mechanotherapy in the treatment of Chronic Periodontitis. *Methods:* A total of 60 patients were selected and divided as group A(Scaling + Metronidazole) and group B (Scaling alone) with 30 patients in each group. A microbiological analysis was carried out to determine the efficacy of these systems in changing the pathogenic flora in deep pockets. Further a multiplex polymerase chain reaction (PCR) was put up to confirm the presence of *Aggregatibacter actinomycetemcomitans* (Aa) *Porphyromonas gingivalis* (Pg) and *Tannerella forsythia* (Tf) in the flora associated with Chronic Periodontitis. *Results:* In group A, there was a clinical improvement which correlated with an improvement in microbiological parameters and these results were maintained 90 days following therapy. In Group B, the microbial flora showed a shift towards baseline at the end of 90 days. *Conclusions:* According to this study the local drug delivery system when used as an adjunct to mechanotherapy resulted in greater improvement in microbiological parameters when compared to mechanotherapy alone.

Key words: Local drug delivery, multiplex PCR, Periodontal pathogens

Introduction

"The only thing that is constant in life is change"
This essence of change has led to constant innovations in the field of medicine, which has evolved consistently from the era of ancient medicine through continuous clinical research into the era where gene therapy and stem cell therapy do not seem impossible dreams. The field of dentistry has also evolved from the era of ancient practices to the present day era of operating microscopes, laser dentistry and tissue engineering. The modernization of dentistry has also given way to changing paradigms, which have evolved the concepts in the field of etiopathogenesis, treatment strategies, non-surgical vs surgical therapy and systemic vs local antibiotic therapy. Dr Max Goodson introduced the concept of local drug delivery based on the theory that improved cellular specificity of a drug will improve the therapeutic index[1]. The local drug delivery devices have evolved from non resorbable to resorbable systems and today we have a variety of agents to choose from[2,3].

Aim of the study

To evaluate the efficacy of a local drug delivery system containing Metronidazole as an adjunct to mechanotherapy in the management of chronic periodontitis.

Materials & methods

The patients were selected from the outpatient department of the Department of dental surgery, Armed Forces Medical College. The criteria for selection were as follows:
Inclusion criteria: i) Age between 30-55 years ii) Minimum 2 sites > 5 mm
Exclusion criteria: i) No h/o any previous periodontal therapy ii) No relevant medical history iii) No h/o antibiotic therapy in last six months.
    Clinical measurements and microbiological samples were taken from the selected sites: (i) Prior to treatment i.e. on day 0 (ii) On day 30 following treatment & (iii) On day 90 following treatment.

Clinical parameters that were considered were: (i) Probing depth (PD)(ii) Clinical attachment level (CAL) (iii) Plaque index (PI) (iv)Gingival index (GI) (v) Gingival bleeding index (GBI).

Microbiological parameters evaluated were: (i) Gram stain (ii) Aerobic culture (iii) Anaerobic culture.

Molecular analysis was done using a multiplex polymerase chain reaction (PCR) [4].After the baseline evaluation, sub gingival scaling was done for all the subjects. The first group (*Group A*) received a single placement of Metronidazole sponge while the second group (*Group B*) received no further treatment. The patients were then asked to report back after 30 days following the treatment in both the groups. All parameters were reassessed and patients were recalled after 60 days for the final 90 day assessment of all parameters.

Results

The clinical and microbiological parameters improved in both the groups at the end of 30 days post therapy. There was reduction in probing depth, gain in clinical attachment and improved scores of all indices (Fig 1,2).

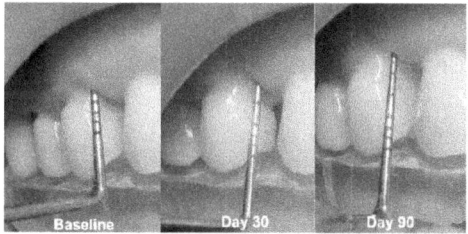
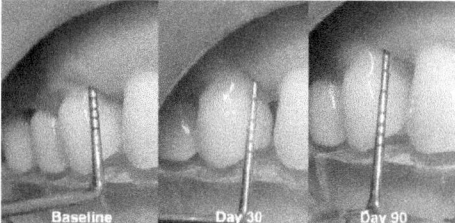

Fig 1: Changes in clinical parameters *(Gp A)*   **Fig 2:** Changes in clinical parameters *(Gp B)*

The microbial parameters showed a reduction in number of colonies seen on cultures as well as a reduction

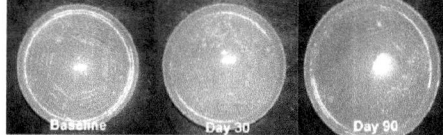
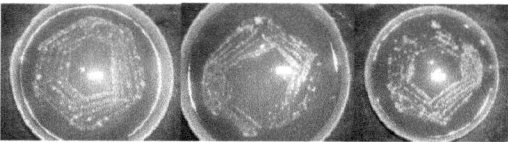

Fig 3: Reduction in no of colonies *(Gp A)*   **Fig 4:** Reversal towards baseline at 90 days *(Gp B)*

Multiplex PCR put up showed the presence of *Porphyromonas gingivalis*(Pg) and *Tannerella forsythia* (Tf) but *Aggregatibacter actinomycetemcomitans* (Aa) could not be detected. However, at the time of assessment at 90 days post therapy though the clinical parameters continued to show improvement in both the groups, there was a significant difference in the microbiological parameters. There was a significant shift towards nonpathogenic gram-positive flora in Group A at 90 days. However in Group B, though the flora showed a shift towards healthy gram-positive flora at the time assessment 30 days post therapy, there was a reversal towards baseline gram-negative pathogenic flora at 90 days post therapy recall. There was a significant difference in both the groups in terms of increase in healthy and decrease in pathogenic flora (Table 1).

Table 1: Comparative changes in healthy and pathogenic flora in both the groups

GROUP	% Increase in Gram positive organism	% Decrease in Gram negative organism
A	44.82	85.68
B	23.91	60.39

Table 2: Comparison between the groups showing the presence of Pg & Tf pre & post therapy

GROUP	*P gingivalis*		*T forsythia*	
	Baseline	**90 days**	**Baseline**	**90 days**
A	6	2	7	1
B	8	5	5	6

The concept of bacterial specificity in the etiopathogenesis of periodontal disease was given by Loesche in 1976 which further evolved to the concept of biofilms and periodontal pathogens like Aa,Pg & Tf. (ref: 1996). Antibiotics have been used to treat periodontal infections routinely as monotherapy as well as adjuncts to the mechanotherapy. The concept of plaque as a biofilm led to the elimination of antibiotics as monotherapy in the management of periodontitis [5]. Today, whenever antibiotics are indicated in the treatment of periodontitis they are almost always used as adjuncts to the conventional mechanotherapy. However, systemic antibiotics have their limitations and this led to the concept of local drug delivery.

Fig 5: LDD (Metronidazole)

The local drug delivery device essentially consists of a drug reservoir and a limiting element, which controls the rate of medicament release. The device used in this study consists of 5% Metronidazole and 95% Collagen.

The antibiotic in the local drug delivery device essentially targets pathogens in three areas: (i) Residing in the periodontal pocket (ii) Exposed cementum /Radicular dentin and (iii) Soft tissue walls of the pocket. The two groups were compared for the following parameters: (i) Changes in clinical parameters (ii) Ability to reduce pathogenic flora. Following sub gingival debridement alone *(Group B)* showed improvement in clinical & microbiological parameters initially (at 30 days evaluation) but there was a reversal towards baseline microbiological picture 90 days following therapy. Following the use of LDD as an adjunct to sub gingival debridement *(Group A)* there was significant reduction in the pathogenic flora and increase in healthy flora. The results were maintained till the end of the study. These findings are in agreement with earlier studies where clinical and microbiological parameters were evaluated[6,7,8,9]

Conclusion

Subgingival debridement will continue to be the mainstay of periodontal therapy. Local drug delivery systems used as adjuncts further help in reduction of pathogenic flora. The decision to use LDD is based on clinical findings, responses to therapy recorded in literature, desired clinical outcomes and patient's dental & medical history. Further studies should include quantitative PCR to quantify changes in the SG flora following treatment, development of chair side evaluation system for microbial analysis, evaluation of multi- drug LDD system and use of nanotechnology for development of LDD systems.

References

[1] Goodson JM, Offenbacher S, Farr DH, Hogan PE. Periodontal disease treatment by local drug delivery. J Periodontol 1985;56:265-72.
[2] *Gary Greenstein and Alan Polson*. The role of local drug delivery in the management of periodontal diseases: A Comprehensive review. *J Periodontol 1998; 69:507-520*
[3] *AAP Position paper*. The role of controlled drug delivery for periodontitis. *J Periodontol 2000; 71:125-140*
[4] Sanz M, Lau L, Herrera D, Morillo JM, Silva A. Methods of detection of *Actinobacillus actinomyctemcomitans, Porphyromonas gingivalis* and *Tannerella forsythensis* in Periodontal microbiology with special emphasis on advanced molecular techniques: A review. J Clin Periodontol 2004; 31:1034-1047.

[5] S S Socransky and AD Haffajee. Bacterial etiology of destructive periodontal disease: Current concepts. *J Periodontol 1992; 63: 322-331*

[6] B.V. Somayaji et al.Evaluation of Antimicrobial Efficacy and Release Pattern of Tetracycline and Metronidazole Using a Local Delivery System.*J Periodontol 1998 69:4, 409-413*

[7] *Ainamo J et al.* Clinical responses to sub gingival application of a metronidazole 25% dental gel compared to the effect of sub gingival scaling in adult periodontitis. *J Clin Periodontol 1992;19:723-9.*

[8] *Klinge B et al.* Three regimens of topical metronidazole compared with sub gingival scaling on periodontal pathology in adults. *J Clin Periodontol 1992; 708-714.*

[9] *Pedrazzoli V, Kilian M, Karring T.* Comparative clinical and microbiological effects of topical sub gingival application of Metronidazole 25% dental gel and scaling in treatment of adult periodontitis. *J Clin Periodontol 1992;19:715-22*

New AMPs from *Staphylococcus* spp., warnerin and hominin, reduce *Staphylococcuccus epidermidis* adhesion and biofilm formation

D. Eroshenko* and **V. Korobov**

Laboratory of microorganisms' biochemical development, Institute of Ecology and Genetics of Microorganisms, Ural Branch of the Russian Academy of Sciences, 13, Goleva, 614081, Perm, Russia.
*Corresponding author: e-mail: dasha.eroshenko@gmail.com , Phone: +79617599718

In view of the high antibiotic-resistance of bacteria in biofilm cationic peptides are becoming more and more promising. The effect of the two peptides warnerin and hominin on the development of *Staphylococcus epidermidis* biofilms was studied. Warnerin caused reduction of the adhesion of all *S. epidermidis* strains, including MRSE, in a dose-dependent manner. The AMPs caused the decrease of adhesion only for live cells in both a polystyrene and glass surface. Immobilized AMPs were more effective than soluble form. The duration of the AMPs inhibitory effect on *S. epidermidis* biofilm was 2 – 5.5 h and depended on *S. epidermidis* strain.

Keywords biofilm; adhesion; staphylococcus; AMPs, warnerin, hominin.

1. Introduction

The formation of bacterial film is one of the major causes of the device-associated infections [1]. The key step in the development of such infections is bacterial adhesion to the implant surface [2]. In this regard, the methods to suppress the earliest stages in the biofilm formation, especially the bacteria-target surface interaction, are becoming very relevant. Low molecular weight cationic peptides represent promising class of antibacterial agents. In this connection it is interesting to study the possibility of their participation in the processes of bacterial adhesion.

2. Materials and Methods

The *Staphylococcus epidermidis* stran GISK 33 (Moscow, Russia) and methicillin-resistant (MRSE) strains ATCC 12228 and ATCC 29887 were used. Bacteria were cultivated in LB medium on a shaker at 37°C to the mid-log phase and twice washed with 0.14 M NaCl (pH 7.2), diluted to 10^7 CFU/ml in appropriated medium and used for experiments. The low molecular weight cationic peptides (AMPs) warnerin (W) and hominin (H) were obtained from the culture media of collection strains *S. warneri* KL-1 and *S. hominis* KLP-1 according to the method described previously [3].

To study the effect of the AMPs on an *S.epidermidis* adhesion to hydrophobic and hydrophilic surfaces the washed live or heat-killed (20 min) *S. epidermidis* GISK 33 cells were suspended in either AMP-free LB medium (control) or LB with 0.5 µg/ml AMPs, then incubated into polystyrene or glass Petri dishes statically at 37°C for 30 min. Then unbounded cells were removed by triplicate washing by PSB (10 mM, pH 7.2). The number of adherent cells was evaluated after their staining with 0.1% gentian violet solution using the direct counting with microvisor "µViso-103" ("Lomo", Russia).

To compare the effect of the soluble and immobilized AMPs on an *S.epidermidis* adhesion the polysterene surfaces were pretreated by 64 µg/ml AMPs during 60 min. Then the washed *S. epidermidis* GISK 33 cells were suspended either in AMP-free LB medium and incubated into surface with immobilized AMPs or in LB with 0.5 µg/ml AMPs and incubated into the bare polystyrene surface. Bacteria were allowed to adsorb for 30 min statically at 37°C without shaking. Then unbounded cells were removed by triplicate washing by PSB (10 mM, pH 7.2). The number of adherent cells was evaluated as described early.

To determinate the duration of anti-biofilm action of AMPs the bacterial cells (10^7 CFU/ml) were suspended in AMPs solutions (1xMIC) and incubated into polystyrene Petri dishes statically at 37°C for 30 min. The peptide-free bacterial suspension was used as the control. Then liquid with unbounded cells was carefully aspirated, adherent cells twice washed by phosphate-saline buffer (10 mM, pH 7.2) or NaCl (0.5 M, pH 7.2), fresh medium LB (2 ml) was added in each Petri dishes with following incubation statically at 37°C for 0, 1, 2, 3, 4 and 24 h. The number of adherent cells was evaluated after their staining with gentian violet (0.1%) using the direct counting.

3. The Results and Discussion

3.1 Effect of warnerin on adhesion of *S. epidermidis* strains

The concentrations of warnerin and hominin required to inhibit the growth of planktonic *S. epidermidis* strains were assessed by the micro-dilution broth method in biofilm promoting medium. MIC values > 8 µg/ml were recorded for all the three strains included in the study (data not shown). The ability of AMPs to inhibit bacterial adhesion was evaluated, as adherent cells number, after incubation of *S. epidermidis* strains for 30 min with different concentrations of the peptides (0.5, 8 and 80 µg/ml). As shown in Figure 1, warnerin was able to reduce the bacterial adhesion of all *S. epidermidis* strains in a dose-dependent manner. The maximal reduction of adhesion (about 80%) was observed at 10xMIC concentrations.

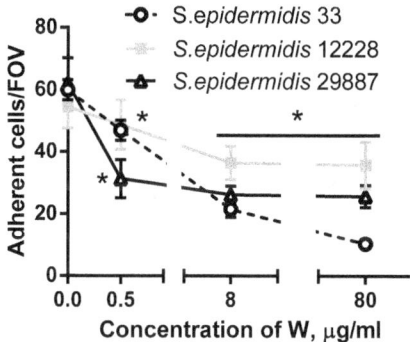

Fig. 1 Adhesion of *S.epidermidis* cells to polystyrene depending on the warnerin (W) concentration in the medium. Data represents the mean ± the 95% CI of at least three independent experiments. * indicates significance compared to control for each strain (P< 0.05)

Only 10xMIC concentrations of warnerin caused reduction (<1 log$_{10}$) in the CFU number during 30 min adhesion (data not shown). However, no evident differences were observed between adhesion methicillin-resistant (ATCC 29887, ATCC 12228) and -sensitive (GISK 33) strains in susceptibility to AMPs. This observation might be relevant for possible future use of warnerin as an antibiofilm agent since MRSE strains offen involved in biofilm-associated infections in clinical settings [4].

3.2 The effect of the AMPs on the adhesion of live and heat-killed cells

Figure 2 shows the warnerin and hominin (0.5 xMIC) reduction the adsorbed cells number. The AMPs caused the decrease of adhesion only for live *S.epidermidis* 33 cells on both a hydrophobic (polystyrene) and a hydrophilic (glass) surface. Isolated from various strains of staphylococci, warnerin and hominin, provided almost the same inhibitory effect on the adhesion to polsterene (48% and 41%, respectively) and glass (46% and 38%, respectively). Both cationic peptides did not affect the adhesion of heat-killed cells.

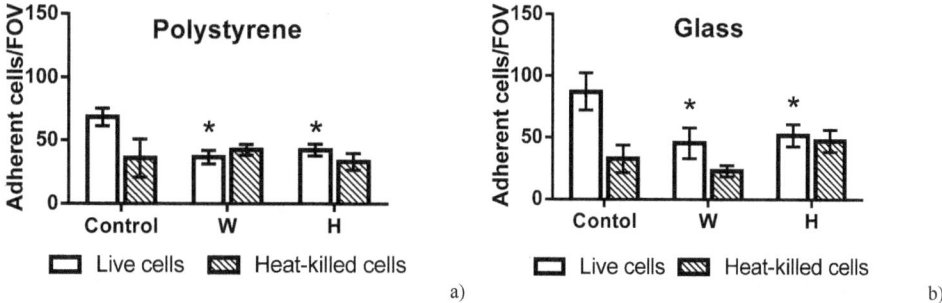

Fig. 2 Action of AMPs (0.5 xMIC) on the adhesion of the live and heat-killed *S.epidermidis* cells to a) hydrophobic (polystyrene) and b) hydrophilic (glass) surfaces. *indicates significance compared to control for each state of cells (P< 0.05).

The reducing of the adherent cells number in the presence of the cationic peptides is not related to their bactericidal effect, since changes in the CFU number for 30 min adhesion are not observed (data not shown). The main difference between living and heat-killed cells is the electro-chemical properties, in particular the zeta potential [5]. Since the living cells adhesion is almost equal to the heat-killed cells adhesion in the peptides presence, probably, the anti-adhesion mechanism of both AMPs action associated with the zeta potential changes.

3.3 The effect of the soluble and immobilized AMPs on the adhesion

Figure 3shows effect of soluble and immobilized AMPs on the S.epidermidis GISK 33 adhesion. Preatreatment of polysterene by warnerin and hominin solutions during 1 h results in the more remarkable reduction of the adsorbed cells number versus soluble form of AMPs. Both the soluble and immobilized AMPs did not change the CFU number for 30 min adhesion (data not shown).

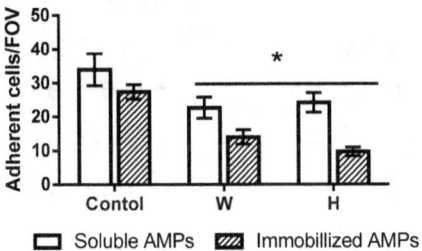

Fig. 3 Effect of soluble and immobilized on the polystyrene AMPs on the S.epidermidis 33 adhesion. *indicates significance compared to control for each state of AMPs (P< 0.05).

There are data about many AMPs that save their activity after their immobilization [6].

3.4 The duration of the AMPs inhibitory effect on S. epidermidis biofilm

The duration of the inhibitory effect was determined after peptide removal. To this end, S. epidermidis cells (33, 12228 and 29887) were allowed to adhere to polystyrene wells in the presence of AMPs (1xMIC for each strain) for 30 min. The peptide and free bacterial cells were then removed by washing two times with PBS or 0.5 M NaCl and fresh LB was added to the Petri dishes. The residual warnerin and hominin anti-biofilm effect was evaluated as the adherent cells number at 2, 4, and 5.5 h post-treatment (Figure 4 and Figure 5).
Interestingly, after AMPs removal, the reduction in the adherent cells number for S. epidermidis 33 was still evident up to incubation for 5.5 h, except the H + NaCl rinsing (Figure 4). Washing by 0.5 M NaCl shortened the generation time by 1.5 and 3.8 times for cells treated with hominin and warnerin respectively and made them almost equal (Table 1).

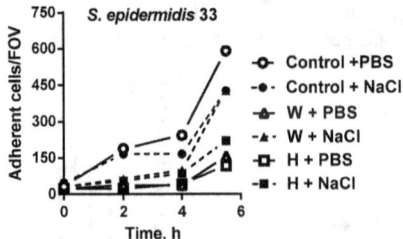

Fig. 4 Duration of the warnerin and hominin inhibitory effect on S. epidermidis GISK 33 biofilm.

The duration of the inhibition of the biofilm formation for MRSE strains (ATCC 29887, ATCC 12228) was shorter compared to methicillin sensitive strain (GISK 33) and lasted at least until 4 h (ATCC 12228) and 2 h (ATCC 29887) post treatment with AMPs (Figure 5 a,b). The adhesion in presence did not alter the subsequent the biomass doubling time, as well as the rinse solution (Table 1).

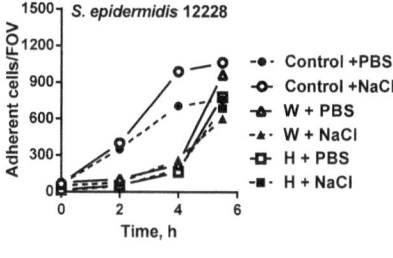
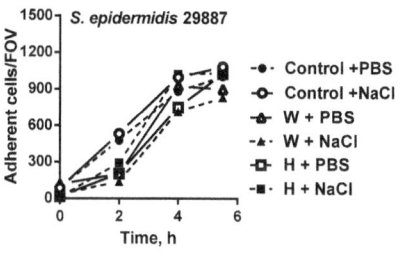

a) b)

Fig. 5 Duration of the warnerin and hominin inhibitory effect on biofilm formation by methicillin-resistant *S. epidermidis* a) ATCC 12228 and b) ATCC 29887 strains.

Table 1 The doubling time (min) of the number of adsorbed bacterial cells after *S. epidermidis* adhesion in presence of AMP (1xMIC) and washing of adherent cells by PBS (10 mM, pH 7.2) or NaCl (0.5 M, pH 7.2). [a] indicates significance compared to control for each strain (P< 0.05), [b] indicates significance in pair PBS-rinsing – NaCl-rinsing (P< 0.05).

Strain	Control +PBS	Control +NaCl	W + PBS	W + NaCl	H + PBS	H + NaCl
33	77±1	77±1	495 ± 34[a]	130 ± 10[a,b]	217 ± 26[a]	141 ± 9[a,b]
12228	60±2	67±1	67±3	70±5	55±4	55±3
29887	67±3	73±2	74±2	79±4	88±17	110±35

No evident effect was observed on CFU number at any of the times tested (data not shown). This suggests that the inhibitory effect of both AMPs on *S. epidermidis* biofilm formation is probably linked to the AMPs ability to regulate the expression of genes that are essential for the development of biofilm [7]. One of the probable mechanisms can be connected with regulation of the agr locus that activates in response to a short autoinducer peptide, AIP [8].

4. Conclusion

The results provide evidence for the low molecular weight cationic peptides warnerin and hominin reduced the bacterial adhesion to abiotic hydrophobic and hydrophilic surface, but the duration of the post-antibacterial effect may differ for *S. epidermidis* strains and in some cases can be eliminated by 0.5M NaCl.

Acknowledgements   The support of RFBR (12-04-01431-a and 14-04-00687) and IRGP the Goverment of Perm Region (C-26/632) is gratefully acknowledged.

References

[1]   Vinh DC and Embil JM. Device-related infections: a review. Journal of long-term effects of medical implants. 2005; 15:467-488.

[2]   Davey ME and O'Toole GA. Microbial biofilms: from ecology to molecular genetics. Microbiology and molecular biology reviews. 2000; 64:847-867.

[3]   Korobov VP, Lemkina LM, Polyudova TV, Akimenko VK. Isolation and Characterization of New Low-Molecular Antibacterial Peptide of the Lantibiotics Family. Microbiology. 2010; 79:206-215.

[4]   Barbier F, Ruppé E, Hernandez D, Lebeaux D, Francois P, Felix B, Ruimy R. Methicillin-resistant coagulase-negative staphylococci in the community: high homology of SCCmec IVa between Staphylococcus epidermidis and major clones of methicillin-resistant Staphylococcus aureus. The Journal of infectious diseases. 2010; 202: 270-281.

[5]   Kłodzińska E, Szumski M, Dziubakiewicz E, Hrynkiewicz K, Skwarek E, Janusz W, Buszewski B. Effect of zeta potential value on bacterial behavior during electrophoretic separation. Electrophoresis. 2010; 31:1590-1596.

[6]   Costa F, Carvalho IF, Montelaro RC, Gomes P, Martins, MCL. Covalent immobilization of antimicrobial peptides (AMPs) onto biomaterial surfaces. Acta biomaterialia. 2011; 7:1431-1440.

[7]   Overhage J, Campisano A, Bains M, Torfs ECW, Rehm BHA, Hancock, REW. Human host defense peptide LL-37 prevents bacterial biofilm formation. Infection and Immunity. 2008; 76:4176-4182.

[8]   Ji G, Beavis R, Novick RP. Bacterial interference caused by autoinducing peptide variants. *Science (New York, N.Y.)*. 1997; 276:2027-2030.

New approach to extend shelf life of Mozzarella cheese using antimicrobial microbes

A. Caridi[*,1], M. L. De Felice[1], A. Piscopo[1], R. Sidari[1], A. Zappia[1] and M. Poiana[1]

[1] Department of AGRARIA, *Mediterranea* University of Reggio Calabria, Via Feo di Vito s/n, 89122 Reggio Calabria, Italy
*Corresponding author: e-mail: acaridi@unirc.it, Phone: +39 0965680578, Fax: +39 0965680578

Mozzarella cheese has a shelf life of approximately 5 to 7 days due to the fast growth of proteolytic *Pseudomonas* strains. *Pseudomonas* can grow in Mozzarella cheese modifying texture and reducing its shelf life. Another factor that reduces its shelf life is the coliform occurrence; effectively, *Escherichia coli* can grow in Mozzarella cheese reducing its safe life and shelf life. Obviously, proteolysis and lipolysis control is of high importance in its preservation. Investigation on shelf life extension of Mozzarella cheese was recently focused on active packaging systems and antimicrobial peptides. However, in spite of the research carried out, at present no completely efficacious method is available to inhibit the fast microbial spoilage of Mozzarella cheese. In our opinion, utilization of antimicrobial microbes can be an innovative strategy to gain this result; so, our aim was to select adjunct cultures of *Lactobacillus* spp. possessing antagonistic activity against both *Pseudomonas* and *E. coli*.

Keywords Mozzarella cheese; shelf life extension; lactic acid bacteria; *Lactobacillus*; *Lactococcus*; *Pseudomonas*; *Escherichia coli*

1. Methodology and results

Ten strains of *Lactobacillus* spp., eight strains of *Escherichia coli*, five strains of *Pseudomonas* spp. and one strain of *Lactococcus* sp. belonging to the collection of our laboratory were used for this research. Among other strains of *Lactobacillus*, in preliminary screenings in Petri plates *Lactobacillus paracasei* ssp. *paracasei* L356 showed antagonistic activity against both *E. coli* (Table 1) and *Pseudomonas* (Table 2).

Table 1 Antagonistic activity of 10 *Lactobacillus* strains against 8 strains of *Escherichia coli*.

Strains of *Lactobacillus*	Strains of *E. coli*								Ref.
	ATCC45922	O26	O157:H7	EC3	EC4	EC5	PP1	PP2	
L. paracasei Lb06	+++	++	++						[3]
L. paracasei Lb07	+++	++	++						[3]
L. plantarum Lb17	++	++	-						[3]
L. plantarum Lb20	-	-	++						[3]
L. plantarum Lb21	+++	++	++						[3]
L. paracasei Lb23	++	+++	+++						[3]
L. paracasei ssp. *paracasei L134*				+++	+++	+++			[1]
L. curvatus L260							+++	+++	[2]
L. paracasei ssp. *paracasei L355*							+++	+++	[2]
L. paracasei ssp. *paracasei L356*							+++	+++	[2]

[1] Caridi A. Selection of *Escherichia coli*-inhibiting strains of *Lactobacillus paracasei* ssp. *paracasei*. Journal of Industrial Microbiology & Biotechnology. 2002; 29:303-8.
[2] Caridi A. Identification and first characterization of lactic acid bacteria isolated from the artisanal ovine cheese Pecorino del Poro. International Journal of Dairy Technology. 2003; 56:105-10.
[3] Geria M, Dambrosio A, Normanno G, Lorusso V, Caridi A. Antagonistic activity of dairy lactobacilli against Gram- foodborne pathogens. Acta Scientiarum. Technology. 2014; 36:1-6.

Accordingly, the strain L356 of *L. paracasei* ssp. *paracasei* was used in co-fermentation (1:1) with a commercial strain of *Lactococcus*, specifically selected to control the fermentation of Mozzarella cheese: strain Lyobac-D, MO 097, Mofin Alce Group. The milk was inoculated with 5% of a preculture in milk of the *Lactococcus* strain and 5% of a preculture in milk of the *Lactobacillus* strain. In Table 3 we compared the performance of the co-fermentation with the standard fermentation, obtained using only the *Lactococcus* strain. The analyses were performed after the Mozzarella cheese production (0 days) and after 7 days of storage at 5°C and at 10°C in governing liquid, inside rigid plastic trays. We investigated Mozzarella cheese for the following physico-chemical and microbiological parameters: aw, pH, titratable acidity (percentage of lactic acid), percentage of dry matter, colour (expressed as L*, a*, and b*), total coliforms, *Pseudomonas*, coccal-shaped lactic acid bacteria, and Lactobacilli. Some of these analyses were also carried out on the governing liquid.

Table 2 Antagonistic activity of 10 strains of *Lactobacillus* against 5 strains of *Pseudomonas*.

Strains of *Lactobacillus*	Strains of *Pseudomonas*				
	LG1	LG2	M1	M2	M3
L. paracasei Lb06	+++	-	-	+++	-
L. paracasei Lb07	+++	-	+++	+++	-
L. plantarum Lb17	+++	+	+++	+	-
L. plantarum Lb20	+++	-	-	-	-
L. plantarum Lb21	+++	-	+	-	-
L. paracasei Lb23	+++	-	+++	+	+
L. paracasei ssp. *paracasei L134*	+++	+	++	+++	+
L. curvatus L260	+++	-	-	-	-
L. paracasei ssp. *paracasei L355*	+++	-	-	-	-
L. paracasei ssp. *paracasei L356*	+++	+++	+++	+++	+

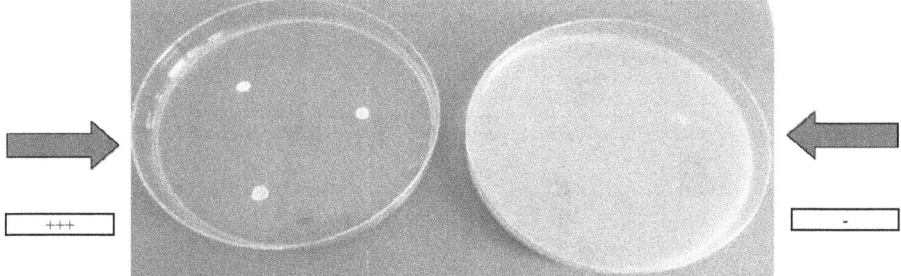

Matching the co-fermentation performance towards the standard fermentation, it is evident: a) a lower pH and a higher lactic acid content after 7 days; b) a higher brightness (L component is higher); c) a higher yellow component in the external part; d) a lower content in *E. coli* and in *Pseudomonas* spp.

2. Discussion

In dairy products, spoilage microorganisms often belong to the Gram-negative group of bacteria; primarily, members of *Enterobacteriaceae* or *Pseudomonas* spp. are frequently detected in milk and dairy products [1]. Spoilage potentials of *Pseudomonas* spp.in cheeses [2] and in Mozzarella cheese [3] were recently investigated. For total coliforms, microbial acceptability limits are considered 10^5 ufc/g of Mozzarella cheese [4]. For *Pseudomonas* spp., microbial acceptability limits are considered 10^6 ufc/g of Mozzarella cheese [5]; according to [6], at this level of contamination the alterations of the product start to appear. In order to control *Pseudomonas* growth, different options were proposed at ICAR2010: a) antimicrobial active packaging [7-9], b) antimicrobial peptides [10-15], c) antimicrobial stress [16], d) bacteriocins produced by lactic acid bacteria [17], e) essential oils [18-26], f) plant, fungal, or algal extracts [27-42]. In order to control *Pseudomonas* growth, different options were proposed also at ICAR2012: a) addition of resveratrol [43], b) addition of sodium lactate [44], c) antimicrobial peptides [45], d) bacteriocins [46], e) essential oils [47-50], f) plant extracts [51-67], g) plant extracts combined with enterocin [68], h) plant extracts combined with honey [69].

Our preliminary results demonstrate that may be useful to employ adjunct cultures of *Lactobacillus* in addition to starter cultures to increase shelf life of Mozzarella cheese. The advantages for dairy industry deriving by the use of this innovative approach may be both economic and environmental. In addition, this approach may modify the current protocols for starter and adjunct culture selection.

Acknowledgements   This work was supported by PRIN 2012: "Long life, high sustainability. To combine the shelf life extension due to a formulation, processing or packaging innovation, with the possible increase of global sustainability of a food product from farm to fork".

Table 3 Physico-chemical and microbiological parameters of Mozzarella cheese and governing liquid.

Mozzarella cheese	co-fermentation			standard fermentation		
	0 days	7 days at 5°C	7 days at 10°C	0 days	7 days at 5°C	7 days at 10°C
a_w	1.00	0.99	1.00	0.99	0.98	0.98
pH	5.54	5.58	5.43	5.71	5.78	5.69
Titratable acidity	0.23	0.34	0.32	0.18	0.20	0.23
% of dry matter	39.26	38.80	38.70	37.62	39.55	39.41
L* - exterior	94.69	95.56	95.56	93.91	93.80	86.34
a* - exterior	-1.18	-0.16	-0.26	-0.29	-0.29	0.07
b* - exterior	7.73	7.75	9.14	7.17	7.23	10.81
L* - interior	90.92	94.66	94.66	93.24	93.90	89.66
a* - interior	-0.03	-0.26	-0.16	-0.23	-0.30	-0.12
b* - interior	11.28	9.46	8.59	8.05	8.19	10.91
Total coliforms	$8.1 \cdot 10^1$	$6.3 \cdot 10^3$	$2.8 \cdot 10^5$	$1.3 \cdot 10^2$	$1.9 \cdot 10^5$	$3.2 \cdot 10^5$
Pseudomonas	$1.4 \cdot 10^2$	$3.8 \cdot 10^5$	$8.6 \cdot 10^5$	$6.0 \cdot 10^3$	$1.3 \cdot 10^6$	$8.3 \cdot 10^6$
Coccal-shaped LAB	$5.8 \cdot 10^{13}$	$3.1 \cdot 10^{11}$	$4.6 \cdot 10^{10}$	$2.8 \cdot 10^{14}$	$1.2 \cdot 10^{14}$	$2.1 \cdot 10^{14}$
Lactobacilli	$2.0 \cdot 10^{11}$	$2.9 \cdot 10^{10}$	$2.1 \cdot 10^{10}$	$4.0 \cdot 10^4$	$3.4 \cdot 10^6$	$4.5 \cdot 10^7$
Governing liquid	0 days	7 days at 5°C	7 days at 10°C	0 days	7 days at 5°C	7 days at 10°C
pH	6.22	5.53	5.10	5.72	5.62	4.91
Titratable acidity	0.27	1.17	2.21	0.19	1.44	1.44
L*	24.09	26.20	36.27	26.55	23.67	33.33
a*	-0.04	-0.09	-0.36	0.11	-0.09	-0.49
b*	-0.87	-1.05	-2.37	-0.88	-1.26	-3.39
Total coliforms	$5.3 \cdot 10^1$	$6.1 \cdot 10^3$	$1.7 \cdot 10^6$	$3.6 \cdot 10^1$	$8.5 \cdot 10^4$	$1.1 \cdot 10^5$
Pseudomonas	$2.0 \cdot 10^1$	$3.9 \cdot 10^6$	$7.6 \cdot 10^6$	$7.0 \cdot 10^2$	$2.4 \cdot 10^7$	$1.2 \cdot 10^8$
Coccal-shaped LAB	$3.4 \cdot 10^7$	$1.1 \cdot 10^7$	$5.7 \cdot 10^8$	$6.1 \cdot 10^6$	$4.4 \cdot 10^6$	$3.3 \cdot 10^7$
Lactobacilli	$1.7 \cdot 10^5$	$4.3 \cdot 10^5$	$2.0 \cdot 10^7$	$9.0 \cdot 10^3$	$4.9 \cdot 10^4$	$1.1 \cdot 10^7$

References

[1]   Domig KJ, Zitz U, Macher S, Kronberger A, Reiter A, Kneifel W. Selective colorimetric detection of Gram-negative re-contaminants in pasteurised milk products by a novel application of the BacT/ALERT 3D system. International Dairy Journal. 2013; 29:21-27.

[2]   Arslan S, Eyi A, Özdemir F. Spoilage potentials and antimicrobial resistance of *Pseudomonas* spp. isolated from cheeses. Journal of Dairy Science. 2011; 94:5851-56.

[3]   Baruzzi F, Lagonigro R, Quintieri L, Morea M, Caputo L. Occurrence of non-lactic acid bacteria populations involved in protein hydrolysis of cold-stored high moisture Mozzarella cheese. Food Microbiology. 2012; 30:37-44.

[4]   European Union. DPR 14/1/1997 n. 54. Regolamento recante attuazione delle direttive 92/46 e 92/47/CEE in materia di produzione e immissione sul mercato di latte e di prodotti a base di latte. European Union, Brussels, Belgium, 1997.

[5]   Bishop JR, White CH. Assessment of dairy product quality and potential shelf life-A review. Journal of Food Protection. 1986; 49:739-53.

[6]   Conte A, Scrocco C, Sinigaglia M, Del Nobile MA. Innovative active packaging systems to prolong the shelf life of Mozzarella cheese. Journal of Dairy Science. 2007; 90:2126-31.

[7]   Ferrocino I, Nasi A, La Storia A, Torrieri E, Scarpa A, Schiavo L, Mauriello G, Ercolini D, Villani F. Antimicrobial packaging affects spoilage microbial populations and volatile organic compounds release in meat stored under vacuum. Abstracts of ICAR2010, p. 200.

[8]   Hauser C, Pischetsrieder M, Wunderlich J, Ziegleder G. Antimicrobial active films. Abstracts of ICAR2010, p. 195.

[9]   La Storia A, Ercolini D, Villani F, Mauriello G. Efficacy of nisin-activated plastic films against meat spoilage bacteria in vitro and in meat. Abstracts of ICAR2010, p. 226.

[10] Beuerman RW, Zhou L, Liu SP, Li J, Pervushin K, Bai Y, Lakshminarayanan R, Tang C, Padmanabhan S, Mavinahalli J, Verma C. Branched peptides with high charge density and enhanced anti-microbial properties. Abstracts of ICAR2010, p. 10.

[11] Mangoni ML, Marcellini L, Zanni E, Uccelletti D, Palleschi C, Barra D. Frog skin antimicrobial peptides promote survival of *Caenorhabditis elegans* infected by a multi-drug resistant strain of *Pseudomonas aeruginosa*. Abstracts of ICAR2010, p. 13.

[12] Neubauerová T, Macková M, Macek T, Šanda M, Králová M, Doležílková I, Holmsgaard PN. Isolation of antimicrobial peptides from plant *Tetragonia tetragonioides*. Abstracts of ICAR2010, p. 20.

[13] Neubauerová T, Macková M, Macek T, Šanda M, Vobůrka Z. Isolation of peptides with antimicrobial activity using diverse purification steps and its sequences determination. Abstracts of ICAR2010, p. 21.

[14] Ohta K, Kajiya M, Zhu T, Nishi H, Mawardi H, Shin J, Elbadawi L, Komatsuzawa H, Kawai T. Synergistic effects of acylated ghrelin on antimicrobial activities mediated by LL-37. Abstracts of ICAR2010, p. 28.

[15] Razquin-Olazarán I, Sánchez-Gómez S, Leiva J, Japelj B, de Tejada GM. Sensitization of *Pseudomonas aeruginosa* to antibiotics by human lactoferricin derived peptides persists several hours after peptide removal. Abstracts of ICAR2010, p. 26.

[16] Sousa AM, Loureiro J, Machado I, Pereira MO. The role of antimicrobial stress on *Pseudomonas aeruginosa* colony morphology diversity, tolerance and virulence. Abstracts of ICAR2010, p. 460.

[17] Stoyanova LG, Ustyugova EA, Sultimova TD, Fedorova GB, Khatruha GS, Netrusov AI. Syntheses of new bacteriocins by different strains *Lactococcus lactis* ssp. *lactis* and their properties. Abstracts of ICAR2010, p. 564.

[18] Adeleye IA, Omadime ME, Daniels FV. Antimicrobial activity of essential oil extracts of *Gongronema latifolium* (Endel.) Decne on bacterial isolates from blood streams of HIV infected patients in Lagos. Abstracts of ICAR2010, p. 98.

[19] Ait-Ouazzou A, Lorán S, Abdelhay A, Laglaoui A, Rota C, Herrera A, Pagán R, Conchello P. Analysis of the antimicrobial activity and chemical composition of *Mentha pulegium*, *Thymus algeriensis* and *Juniperus phoenicea* essential oils from Morocco. Abstracts of ICAR2010, p. 82.

[20] Ait-Ouazzou A, Cherrat L, Somolinos M, Lorán S, Rota C, Pagán R. Antimicrobial activity of four hydrocarbon monoterpenes acting alone or in combination with heat. Abstracts of ICAR2010, p. 99.

[21] Espina L, Ait-Ouazzou A, Somolinos M, Lorán S, Conchello P, García D, Pagán R. Chemical composition of commercial citrus fruit essential oils and evaluation of their antimicrobial activity acting alone or in combination with heat. Abstracts of ICAR2010, p. 122.

[22] Ferreira S, Costa B, Domingues FC. Antimicrobial activity of *Coriander* oil. Abstracts of ICAR2010, p. 97.

[23] Osho A, Timothy A, Scott O. Antimicrobial effect of the essential oils of some rare Nigerian medicinal plants. Abstracts of ICAR2010, p. 110.

[24] Patel MP, Parabia FM, Doshi HV. Identification of antibacterial compounds from the essential oil of seven Indian medicinal plants. Abstracts of ICAR2010, p. 143.

[25] Santos GKN, Dutra KA, Lira DD, Navarro DMAF, Câmara CG, Gusmão NB. Antimicrobial activity of *Alpinia purpurata* (Viell) Schum essential oil. Abstracts of ICAR2010, p. 94.

[26] Sultanbawa Y, Currie M, Cusack A, Chaliha M. Antimicrobial activity of Australian native plant essential oils against food-related bacteria. Abstracts of ICAR2010, pp. 95-96.

[27] Adeshina GO, Onaolapo JA, Odama LE, Ehinmidu JO. Antimicrobial activity of hexane fraction of *Alchornea cordifolia* leaf. Abstracts of ICAR2010, p. 100.

[28] Belyagoubi L, Abdelali KN, Khelifi A, Janakat S. Antibacterial activity of aqueous extracts of three Algerian desert truffles against *Pseudomonas aeruginosa* and *Staphylococcus aureus* in vitro. Abstracts of ICAR2010, p. 87.

[29] Belyagoubi L, Chaibi R, Aissaoui FZ, Benamar ZEO. Evaluation of antimicrobial activity of seaweed *Spirogyra* sp. from Algeria. Abstracts of ICAR2010, p. 132.

[30] Gupta S, Jaiswal AK, Abu-Ghannam N. Modelling the growth inhibition of common food spoilage and pathogenic microorganisms in presence of solvent extract from Irish York cabbage. Abstracts of ICAR2010, p. 165.

[31] Hassan LEA, Koko WS, Yagi SMA. In vitro antimicrobial activities of chloroform, hexane and ethanolic extracts of *Citrullus lanatus* var. *citroides* (Wild melon). Abstracts of ICAR2010, p. 148.

[32] Kavitha HU, Satish S. Isolation and identification of antibacterial active compound from *Carum copticum* L. Abstracts of ICAR2010, p. 161.

[33] Mahajan DC, Tatke P, Naharwar V. Evaluation of antimicrobial activity of seeds of *Brassica juncea* and *Brassica nigra*. Abstracts of ICAR2010, p. 133.

[34] Marandino A, de Martino L, de Feo V, Coppola R, Nazzaro F. Antimicrobial and anti quorum sensing activities of extracts from *Hypericum connatum*. Abstracts of ICAR2010, p. 103.

[35] Medina E, de Castro A, Romero C, Brenes M. Antimicrobial substances from olive products: implications on health, food, and agriculture of glutaraldehyde-like compounds. Abstracts of ICAR2010, p. 114.

[36] Nazzaro F, Fratianni F, Riccardi R, Spigno P, Coppola R. Antimicrobial and anti quorum sensing activities of two typical *Brassica* cultivars present in the Campania region (Southern Italy). Abstracts of ICAR2010, p. 104.

[37] Ogundare AO. Phytochemical and antibacterial properties of *Combretum mucronatum* leaf extract. Abstracts of ICAR2010, p. 169.

[38] Ogundare AO. Phytochemical and antibacterial properties of the leaf extracts of some edible plants in Nigeria: *Vernonia amygdalina*, *Ocimum gratissimum*, *Corchorous olitorious* and *Manihot palmate*. Abstracts of ICAR2010, p. 170.

[39] Pandey AK, Mishra A, Mishra AK, Bhargava A. Anti-infective and free radical scavenging activities of phytochemicals present in some plant extracts. Abstracts of ICAR2010, p. 84.

[40] Saavedra MJ, Borges A, Dias C, Aires A, Bennett RN, Rosa ES, Simões M. Antimicrobial activity of phytochemicals and their synergy with streptomycin against pathogenic bacteria. Abstracts of ICAR2010, p. 363.

[41] Stefanovic O, Comic Lj, Radojevic I, Stankovic M. Antibacterial activity of *Cytisus nigricans* L. extracts and their synergistic interaction with antibiotics. Abstracts of ICAR2010, p. 88.

[42] Vieira L, Ferreira S, Mendonça AJ, Mendonça DI, Domingues F. Antimicrobial properties of the *Eragrostis viscosa* extracts. Abstracts of ICAR2010, p. 111.

[43] Martínez-González O, Azcue LE, Martín MSV, de Vega Castaño MC, Valencia CC, Egea JS. Addition of resveratrol to smoked sea bass (*Dicentrarchus labrax*) fillets in order to achieve a longer shelf life. Abstracts of ICAR2012, p. 7.

[44] Colak H, Bingol EB, Cetin O, Hampikyan H. Effects of sodium lactate on the shelf life of Cig Kofte - A Turkish Traditional Raw Meatball. Abstracts of ICAR2012, p. 198.

[45] Quintieri L, Caputo L, Morea M, Baruzzi F. Control of Mozzarella spoilage bacteria by using bovine lactoferrin pepsin-digested hydrolysate. Abstracts of ICAR2012, p. 17.

[46] Bouktit N, Stocker P, Fons M, Benallaoua S. Identification and purification of a novel unmodified bacteriocin from an Algerian extremely halotolerant *Oceanobacillus* sp. strain. Abstracts of ICAR2012, p. 25.

[47] Alves-Silva JM, dos Santos SMD, Pintado MM, Pérez-Álvarez JA, Martos MV, Fernandez-López J. In vitro antimicrobial properties of coriander (*Coriandrum sativum*) and parsely (*Petroselinum crispum*) essential oils encapsulated in β-cyclodextrin. Abstracts of ICAR2012, p. 222.

[48] Chaftar N, Girardot M, Bergès T, Labanowski J, Hani K, Ghrairi T, Frere J, Imbert C. Antimicrobial screening of 20 essential oils; interest of essential oils extracted from Tunisian plants. Abstracts of ICAR2012, p. 174.

[49] El Beyrouthy M, Landolsi R, Darwish M, Wakim LH, Jaoudeh CA, Dhifi W, Mnif W, Arnold N, Bouez J. Antimicrobial activity and chemical composition of the essential oil of *Satureja thymbra* from Lebanon and comparison with its major components. Abstracts of ICAR2012, p. 166.

[50] Fernandes Júnior A, Machado BFMT, Barbosa LN, Albano M, Stanuch M. Vapor-phase: antimicrobial activities of essential oils. Abstracts of ICAR2012, p. 255.

[51] Abdel-Rahim NA, Fadl-Elmula IM, Almagboul AZ. In vitro antimicrobial activity of *Moringa oleifera* Lam. Abstracts of ICAR2012, p. 220.

[52] Apenteng JA, Agyare C, Adu F, Boakye YD. Antimicrobial activities of different leaf extracts of *Pupalia lappacea*. Abstracts of ICAR2012, p. 164.

[53] Bandh SA, Lone BA, Kamili AN, Ganai BA, Chishti MZ, Bhat FA. In vitro antimicrobial activity of three extracts of *Nepetacataria*. Abstracts of ICAR2012, p. 221.

[54] Cardoso F, Matos O. Screening of methanol Portuguese ethno botanical plant extracts for the antimicrobial activity. Abstracts of ICAR2012, p. 243.

[55] De Vasconcelos MCBM, Ferreira-Cardoso JV, Simões M, Rosa EA, Saavedra MJ. In vitro antimicrobial activity of chestnut fruit and olive leaf residues against *Pseudomonas aeruginosa* and *Staphylococcus aureus*. Abstracts of ICAR2012, p. 216.

[56] Djahra AB, Bordjiba O, Benkherara S, Benkaddour M. The in-vitro evaluation of antibacterial properties of *Marrubium vulgare* L. flavonoids grown in Algeria. Abstracts of ICAR2012, p. 251.

[57] Hernández-Castillo FD, Castillo-Reyes F, García-Gonzalez A, de Rodriguez DJ, Rodríguez-Herrera R, Aguilar-Gonzalez CN. Antimicrobial effect of polyphenols from semi-desert Mexican plants on plant pathogenic bacteria. Abstracts of ICAR2012, p. 546.

[58] Hummelová J, Rondevaldová J, Kokoška L. Antimicrobial activity of selected isoflavones. Abstracts of ICAR2012, p. 171.

[59] Lone BA, Bandh SA, Chishti MZ, Bhat FA, Tak H. In vitro antimicrobial activity of methanol extract of *Euphorbia heliscopia*. Abstracts of ICAR2012, p. 219.

[60] Nassirabady N. Antimicrobial effects of *Avicennia alba* on a number of pathogenic bacteria. Abstracts of ICAR2012, p. 547.

[61] Nazzaro F, Fratianni F, Ombra MN, Maione M, Coppola R. Antimicrobial activity of purple basil (*Ocimum basilicum* var. *purpurascens*). Abstracts of ICAR2012, p. 169.

[62] Nissa H, Kamili AN, Shajr-ul-Amin, Bandh SA, Lone BA. Antimicrobial and antioxidant activity of alcoholic extracts of *Rumex dentatus* L.. Abstracts of ICAR2012, p. 172.

[63] Pereira V, Vasconcelos MC, Dias C, Aires A, Carvalho E, Castro P, Rosa E, Saavedra MJ. In vitro antimicrobial activity of *Eucalyptus globulus* leaf extracts against 16 *Pseudomonas aeruginosa* isolates. Abstracts of ICAR2012, p. 218.

[64] Saidana D, Eddine HK, Boussadia O, Osmane N, Mariem FB, Daami M., Braham M. Antimicrobial activity of olive (*Olea europaea* L. Cv. *Chemlali*) aerial part. Abstracts of ICAR2012, p. 168.

[65] Silva LN, Trentin DS, Zimmer KR, Treter J, Brandelli CLC, Frasson AP, Tasca T, Silva AG, Silva MV, Macedo AJ. Medicinal plants from Brazilian Caatinga: antibiofilm activity, cytotoxicity evaluation and phytochemical screening. Abstracts of ICAR2012, p. 231.

[66] Vasić S, Radojević I, Stanković M, Stefanović O, Ćomić Lj, Topuzović M, Nikolić M. Exploring antibacterial and antifungal activity of horehound, *Marrubium peregrinum* L. extracts. Abstracts of ICAR2012, p. 206.

[67] Zveryachenko TS, Zveryachenko AS. New antimicrobial and anti-inflammatory remedies from balsam poplar. Abstracts of ICAR2012, p. 235.

[68] García JD, Baños A, Núñez C, Abad P, Guillamon E, Martínez-Bueno M, Maqueda M, Valdivia E. Innovative air-sanitizing system for the food industry based on vegetable extracts combined with enterocin AS-48. Abstracts of ICAR2012, p. 83.

[69] Moussa A, Noureddine D, Saad A, Abdelmalek M. Anti-*Pseudomonal* activity of pure honey alone and in combination with *Nigella sativa* seeds. Abstracts of ICAR2012, p. 152.

Novel $bla_{CTX-M-2}$-type gene coding extended spectrum beta-lactamase CTX-M-115 discovered in nosocomial *Acinetobacter baumannii* isolates in Russia

I. Dyatlov[1], E. Astashkin[1], N. Kartsev[1], O. Ershova[2], E. Svetoch[1], V. Firstova[1], N. Fursova[*,1]

[1]State Research Center for Applied Microbiology and Biotechnology, 142279 Obolensk, Moscow region, Russia
[2]Burdenko Neurosurgery Institute, 16 4-th Tverskaya-Yamskaya street, 125047 Moscow, Russia
*Corresponding author: e-mail: n-fursova@yandex.ru, Phone: +74967360079

Keywords: *Acinetobacter baumannii*; ESBLs; CTX-M-2-type beta-lactamase; $bla_{CTX-M-115}$ gene; IncA/C plasmid

1. Introduction

Nosocomial infections are one of the most important problems of modern healthcare worldwide. One of the leading agents of these infections in the intensive care unit (ICU) is *Acinetobacter baumannii*. This pathogen is often characterized by multi- (MDR), extreme- (XDR) or pan-resistance (PDR) to antimicrobials [1]. Major mechanism of *A. baumannii* resistance to beta-lactams is producing of OXA-type beta-lactamases and extended spectrum beta-lactamases (ESBLs) including CTX-M-type ESBLs [2]. The CTX-M-beta-lactamase lineage exhibits a striking plasticity, with a large number of allelic variants belonging to at least six sublineages or groups (CTX-M-1, CTX-M-2, CTX-M-8, CTX-M-9, CTX-M-25 and KLUC) which derived from the genomes of *Kluyvera* spp. [3]. The cefotaximase CTX-M-2 gene has been identified in *Salmonella enterica* subsp. *enterica* serovar Typhimurium in 1996 by Bauernfeind et al. [4] as an ancestor of a new subgroup of plasmidic class A beta-lactamases. Since that time at least 20 $bla_{CTX-M-2}$ alleles were identified [3]. Mechanism of a new ESBL gene generation was recognized – accumulation of point mutations on the gene sequence [3]. In this study we described a new point mutation resulting in appearance of a new $bla_{CTX-M-2}$ type gene.

2. Materials and Methods

2.1. Bacterial strains

A. baumannii isolates (n=59) resistant to a wide spectrum of antibiotics were collected from a neurosurgical ICU in Moscow hospital from January, 2013 to May, 2014. Bacterial identification was done by Vitek-2 (Biomerieux, France) and MALDI-TOF Biotyper (Bruker Daltonik GmbH, Germany). Bacteria were grown on Muller-Hinton media (HiMedia, India) at 37 °C and stored at -70 °C until analysis. *A. baumannii* nosocomial strains B-5 and B-120 carrying a novel $bla_{CTX-M-115}$ gene were isolated from the endotracheal aspirate of a patient on mechanical ventilation in Jan and Feb, 2013.

2.2. Antibacterial susceptibility

Minimal inhibitory concentrations (MICs) of antibacterials for *A. baumannii* isolates were determined by Vitek-2 device (Biomerieux, France). Results were interpreted according to the 2014 European Committee on Antimicrobial Susceptibility Testing recommendations (http://www.eucast.org/clinical_breakpoints/). *E. coli* strains ATCC 25922 and ATCC 35218 were used for quality control.

2.3. PCR-detection

PCR was performed using previously described oligonucleotide primers to detect bla_{TEM}, bla_{CTX-M}, bla_{OXA}-type beta-lactamase genes [5, 6, 7, 8]. In addition, class 1 and 2 integrons were detected using primers for the integrase gene and variable region of integrons [7]. The *adeR* regulatory gene of efflux gene cluster AdeABC was detected [8], and *ompA* gene coding porine protein was identified [9]. Bacterial lysates were prepared from an overnight culture by heating at 98 °C [7]. The PCR reaction was carried out in a 25 μl reaction mixture using GradientPalmCycler (Corbert Research, Australia). Amplification products were analysed by electrophoresis in 1% agarose gel in a Sub-Cell GT apparatus (BioRad, US).

2.4. DNA sequencing analysis

Cycle sequencing reactions were performed using the ABI PRISM® BigDye™ Terminator v. 3.1 kit. Purified products were analyzed on an ABI PRISM 3100-Avant automated DNA sequencer in SINTOL center for collective use (Moscow, Russia).

2.5. Bioinformatic analysis

A computer analysis of available gene sequences was performed using Vector NTI9 software (Invitrogen, USA). Nucleotide aligments were done using BLAST web resourse (http://blast.ncbi.nlm.nih.gov/Blast.cgi). Sequenced beta-lactamase genes have been submitted into GenBank database and Lahey database (www.lahey.org/Studies).

2.6. GenBank accession number

The DNA sequence data for the novel $bla_{CTX-M-115}$ gene and adjacent genetic environment sequences were assigned accession numbers KJ911020 and KJ911021 in the EMBL/GenBank database.

3. Results and Discussion

3.1 Strain sources and antibacterial susceptibility

Analyses of the etiological structure of the nosocomial infections in neurosurgical ICU in Moscow in 2013-2014 show that 18% of cases are caused by *A. baumannii*. A total of 59 nosocomial *A. baumannii* isolates resistant to a wide spectrum of antibiotics have been collected, among them 75% from the respiratory system, 10% from liquor, 8% from urine, and 7% from blood. The isolates were resistant to amoxicillin/clavulanate (100%), cefotaxime (100%), ceftazidime (100%), ceftriaxone (100%), ciprofloxacin (100%), nitrofurantoin (100%), cefepime (98%), chloramphenicol (98%), meropenem (96%), imipenem (93%), amikacin (86%), gentamicin (82%), trimethoprim (82%), co-trimoxazole (82%), ampicillin/sulbactam (73%), tobramycin (67%), tetracycline (57%), cefoperazone/sulbactam (34%), and tigecycline (10%) (Fig.1). So, analysis of antibacterial susceptibility show that major (70%) nosocomial strains are MDR and XDR bacteria. Only tigecycline can be regarded as safe drug in this situation.

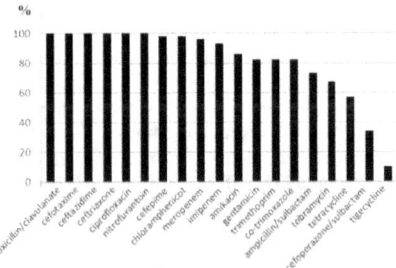

Fig. 1   Percentage of *A. baumannii* isolates resistant to antibacterials

3.2 PCR-detection and sequencing of the resistance genes

It was shown that poly-resistance of the *A. baumannii* isolates under study was based on the molecular mechanisms involving the presence of the resistance determinants. It was detected by PCR that 54% of isolates had class 1 integrons carrying gene cassettes *dfrA17-aadA5* (GenBank: KJ579283) and *aacC1-orfX-orfY-aadA1* (GenBank: KM009103; KM009104; KM009105). Class 2 integron was detected in one isolate only and carried the gene cassette *dfrA1-sat2-aadA1* (GenBank: KM009107). The *adeR* gene coding the regulator of the efflux gene cluster AdeABC was detected in 98% *A. baumannii* isolates (KJ363325), and *ompA* gene coding porin OmpA was identified in 95% isolates under study (KJ363323). Beta-lactamase bla_{OXA}-type genes were detected in all 59 *A. baumannii* isolates: 100% of them carrying bla_{OXA-51}-like genes (GenBank: KJ187467; KJ187469; KJ187471; KJ187473), 73% isolates carrying bla_{OXA-40}-like genes (GenBank: KJ187468; KJ187470), and 14% isolates carrying bla_{OXA-23}-like genes (GenBank: KJ187472; KJ187474). Some isolates (7%) have genes coding TEM-type beta-lactamases, and 18% isolates – genes coding CTX-M-type ESBLs (Fig. 2). Three different alleles of bla_{CTX-M}-type genes were identified by sequencing analysis: $bla_{CTX-M-2}$ (GenBank: KJ187478;

KM085434), $bla_{CTX-M-15}$ (GenBank: KF971880), and a novel allele $bla_{CTX-M-115}$ submitted to Lahey Clinic database (http://www.lahey.org/Studies/other.asp#table1) and to GenBank (KJ911020; KJ911021). A novel variant of $bla_{CTX-M-2}$-like gene was detected in two clinical isolates *A. baumannii* B-5 and B-120 collected from the same patient with mechanical ventilation in Jan and Feb 2013.

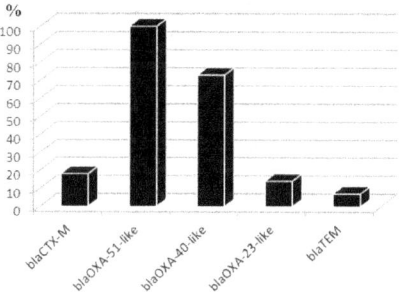

Fig. 2 Prevalence of beta-lactamase genes among *A. baumannii* nosocomial isolates

3.3 Sequence analysis of novel $bla_{CTX-M-115}$ gene

Sequencing of the $bla_{CTX-M-115}$ gene and adjacent genetic environments was done using the set of specific primers ISEcp1U1 [3] and CTX-MR1 [4], and the primers designed in this study CTX-M2-K-F1 (5'-tgaaaaatctgacgctgggt-3') and CTX-M2-K-R (5'-gcaagacaagactgaagttcagg-3') (Fig. 3). Comparison of the $bla_{CTX-M-115}$ gene primary structure with the $bla_{CTX-M-2}$ gene primary structure (GenBank: X92507) revealed six nucleotide substitutions: G336T, G357T, T441G, G751A, A835G, and G868A, resulting in three amino acid substitutions in enzyme molecule, namely Val251Ile, Ile279Val, and Gly290Ser. So, significant nucleotide substitutions are localized on the 3' end of the $bla_{CTX-M-115}$ gene. Similarity search using BLAST software (http://blast.ncbi.nlm.nih.gov) determined that the most similar sequences are *Kluyvera ascorbata* bla_{kluA-1} gene (GenBank: AJ272538.2) and *Escherichia coli* $bla_{CTX-M-124}$ gene (GenBank: JQ429324.1). Our sequence differed from those by one nucleotide substitution G868A resulting in an amino acid substitution Gly290Ser. It was shown that the $bla_{CTX-M-115}$ gene is associated with the ~60 Kb-plasmid belonged to IncA/C incompatibility group [10]. The adjacent genetic environment of $bla_{CTX-M-115}$ gene includes the 3'-end of the mobile genetic element ISEcp1 and the 43 bp-intergenic spacer upstream; and the 59 bp-nucleotide sequence which has 100% identity to the sequences of enterobacterial plasmids downstream (Fig. 3). The presence of ISEcp1 upstream and 59-bp fragment downstream of $bla_{CTX-M-115}$ gene shows that *A. baumannii* could get the gene from enterobacteria. Genetic transfer between *Enterobacteriaceae* and *A. baumannii* through the plasmid carrying $bla_{CTX-M-2}$ gene as part of the complex class 1 integron In35 was described by Ramírez et al. [11].

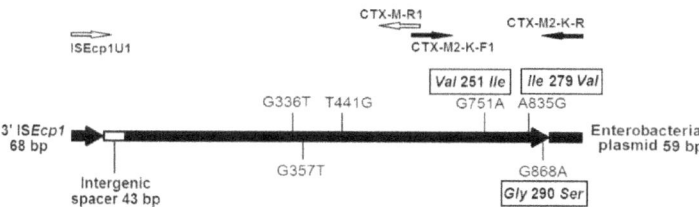

Fig. 3 The $bla_{CTX-M-115}$ gene (876 bp), adjacent genetic environments, localization of the specific nucleotide substitutions and oligonucleotide primers

4. Conclusion

A. baumannii strains collected from the patients of neurosurgical ICU in Moscow hospital in 2013-2014 are causative agents of nosocomial infections characterized by MDR and XDR phenotypes. Molecular mechanism of this phenomenon is presence of the resistance genes on the bacterial genomes: bla_{CTX-M}-type, bla_{TEM}-type, bla_{OXA}-type, class 1 and class 2 integrons, *ade* efflux pump, and *ompA* porin genes. A novel $bla_{CTX-M-115}$ gene encoding CTX-M-2-type beta-lactamase in two MDR nosocomial *A. baumannii* isolates collected in the neurosurgical ICU in Moscow was identified in the study. This fact is important for the research of molecular mechanisms of antibacterial resistance, evaluation of the epidemiological situation, and the choice of optimal therapy strategies in the future.

Acknowledgements   The study was done on the frame of Russian R&D#037 «Improved methods for the identification and studying of biological, molecular-genetic, and biochemical characteristics for causative agents of clostridiosis, legionellosis, listeriosis, etc. bacterial food-borne infections including atypical bacteria».

References

[1]   Magiorakos AP1, Srinivasan A, Carey RB, Carmeli Y, Falagas ME, Giske CG, Harbarth S, Hindler JF, Kahlmeter G, Olsson-Liljequist B, Paterson DL, Rice LB, Stelling J, Struelens MJ, Vatopoulos A, Weber JT, Monnet DL. Multidrug-resistant, extensively drug-resistant and pandrug-resistant bacteria: an international expert proposal for interim standard definitions for acquired resistance. Clin Microbiol Infect. 2012;18(3):268-81.

[2]   Durante-Mangoni E1, Zarrilli R.Murray PR, Rosenthal KS, Kobayashi GS, Pfaller MA. Medical microbiology. 4th ed. St. Louis: Mosby; 2002. Global spread of drug-resistant Acinetobacter baumannii: molecular epidemiology and management of antimicrobial resistance. Future Microbiol. 2011;6(4):407-22.

[3]   D'Andrea MM, Arena F, Pallecchi L, Rossolini GM. CTX-M-type β-lactamases: a successful story of antibiotic resistance. Int J Med Microbiol. 2013;303(6-7):305-17.

[4]   Bauernfeind A, Stemplinger I, Jungwirth R, Ernst S, Casellas JM. Sequences of beta-lactamase genes encoding CTX-M-1 (MEN-1) and CTX-M-2 and relationship of their amino acid sequences with those of other beta-lactamases. Antimicrob Agents Chemother. 1996;40(2):509-13.

[5]   Eckert C, Gautier V, Arlet G. DNA sequence analysis of the genetic environment of various blaCTX-M genes. J Antimicrob Chemother. 2006;57(1):14-23.

[6]   Edelstein M1, Pimkin M, Palagin I, Edelstein I, Stratchounski L. Prevalence and molecular epidemiology of CTX-M extended-spectrum beta-lactamase-producing Escherichia coli and Klebsiella pneumoniae in Russian hospitals. Antimicrob Agents Chemother. 2003 Dec;47(12):3724-32.

[7]   Priamchuk SD, Fursova NK, Abaev IV, Kovalev IuN, Shishkova NA, Pecherskikh EI, Korobova OV, Astashkin EI, Pachkunov DM, Kruglov AN, Ivanov DV, Sidorenko SV, Svetoch EA, Diatlov IA. [Genetic determinants of antibacterial resistance among nosocomial Escherichia coli, Klebsiella spp., and Enterobacter spp. isolates collected in Russia within 2003-2007]. Antibiot Khimioter. 2010;55(9-10):3-10. Russian.

[8]   Hujer KM, Hujer AM, Hulten EA, Bajaksouzian S, Adams JM, Donskey CJ, Ecker DJ, Massire C, Eshoo MW, Sampath R, Thomson JM, Rather PN, Craft DW, Fishbain JT, Ewell AJ, Jacobs MR, Paterson DL, Bonomo RA. Analysis of antibiotic resistance genes in multidrug-resistant *Acinetobacter* sp. isolates from military and civilian patients treated at the Walter Reed Army Medical Center. Antimicrob Agents Chemother. 2006;50(12):4114-23.

[9]   Bratu S1, Landman D, Martin DA, Georgescu C, Quale J. Correlation of antimicrobial resistance with beta-lactamases, the OmpA-like porin, and efflux pumps in clinical isolates of *Acinetobacter baumannii* endemic to New York City. Antimicrob Agents Chemother. 2008;52(9):2999-3005.

[10] Carattoli A, Bertini A, Villa L, Falbo V, Hopkins KL, Threlfall EJ. Identification of plasmids by PCR-based replicon typing. J Microbiol Methods. 2005;63(3):219-28.

[11] Ramírez MS, Merkier AK, Quiroga MP, Centrón D. *Acinetobacter baumannii* is able to gain and maintain a plasmid harbouring In35 found in Enterobacteriaceae isolates from Argentina. Curr Microbiol. 2012;64(3):211-3.

Our Experience in Managing Severe Intra-abdominal Infections in an Intensive Care Unit of a Tertiary Care Hospital, 2011-2014

Pérez Civantos D[*,1], Martínez García P[*], Farje Mallqui V[*], Jerez Gómez-Coronado V[*], Robles Marcos M[*], Fajardo Olivares M[&], Sánchez Vega J[¥], Fariñas Seijas H[¥]

*Department of Intensive Care Unit. University Hospital Infanta Cristina. University of Extremadura. Badajoz. Spain.

[&] Department of Microbiology. University Hospital Infanta Cristina. University of Extremadura. Badajoz. Spain.

[¥] Center for Clinical Research of Badajoz Area (CICAB). Spain. University Hospital Infanta Cristina University of Extremadura. Badajoz. Spain.

[1] Main author: e-mail: dpcivantos@gmail.com; contact details: Demetrio Pérez Civantos. C/ Padre Rafael 2-7ºB. 06002. Badajoz. Spain. Telephone +34924247769. Mobile: +34656532450; Hospital Adress: University Hospital Infanta Cristina. Carretera de Portugal s/n. 06800. Badajoz. Spain.

Key words: Intra-abdominal infection. Peritonitis. Abdominal sepsis.

Introduction:

Peritonitis is defined as an inflammatory process caused by any irritant substance or any microorganism as bacteria, fungus or virus. Intra-abdominal infection is defined as peritonitis cause by infectious pathogens. Finally, abdominal sepsis includes systemic manifestations of a severe peritoneal infectious inflammation. Peritonitis can be classified as community or nosocomial.

Intra-abdominal infections are the second most common cause of infectious mortality in intensive care units (ICU)[1]. Complicated intra-abdominal infections (cIAI) extend beyond the hollow viscous of origin into the peritoneal space and are associated with local or diffuse peritonitis[1,2]. They are an important cause of morbidity and are frequently associated with poor prognosis [1,2]. However, an early clinical diagnosis, followed by adequate surgery and prompt initiation of appropriate antimicrobial therapy can limit the associated mortality[3]. A few enteric species belonging to the normal gut flora are involved in intraperitoneal infections[4]. Healthcare associated (HA) disease has been related to the presence of more resistant flora, such as *Pseudomonas aeruginosa, Enterobacter* spp, *Enterococci,* and *Candida* spp[5]. The outcome of cIAI depends on several risk factors, such as advanced age, comorbidity and underlying malignancy, severity of illness, degree of peritoneal involvement, and presence of a healthcare environment. Recent guidelines issued by the Infectious Diseases Society of America (IDSA) and the Surgical Infection Society (SIS) recommend broad-spectrum coverage in the case of high-risk peritonitis, i.e., healthcare-associated disease or severe community-acquired (CA) cIAI, as the consequences of treatment failure may be more significant. However, this hypothesis has not been rigorously examined in clinical trials[6].

The aim of our study was to know the etiology and outcome of severe peritonitis at our institution, in order to validate the 2010 IDSA guidelines concerning its management. We also tried to confirm the real incidence of shock in both groups of cIAI and to confirm if correct empiric treatment and early surgery have impact on ICU mortality.

Methods:

This is a retrospective observational and longitudinal study conducted at the ICU of the University Hospital Infanta Cristina, Badajoz, Spain, which is a tertiary university hospital. With the approval of the Clinical Research and Ethical Committee (CREC), all adult patients admitted to the ICU with severe intra-abdominal infection who underwent any type of abdominal surgery procedure between January 1st, 2011 and September 30th, 2014 were included. A total of 57 patients were included in the study.

Patients were divided into two groups, health-care acquired infection (HA) group A and community acquired (CA), group B.

For all patients the following data were collected: peritoneal samples for microbiological study, demographic data; immune status; time to surgery (hours); comorbidity (Charlson index); place of illness onset (community or hospital); anatomic origin of the peritonitis (stomach-small bowel, hepatobiliary and pancreas; colonic); Mannheim Index; severity scores at admission (APACHE II and SOFA); presence of septic shock, days on mechanical ventilation (MV), length of stay in ICU, use of antimicrobial drugs fifteen days before the surgery; microbial isolation, adequacy of empiric treatment, change of antibiotic, relaparotomy, open abdomen or vacuum, mortality in ICU and at 30 days of ICU discharge.

Spontaneous bacterial peritonitis and peritonitis related to continuous ambulatory peritoneal dialysis were excluded from the analysis.

Definitions:

Only samples of peritoneal fluid obtained by sterile puncture in theatre were included. Peritoneal fluid was considered positive if at least one organism was isolated.

Complicated intra-abdominal infection (cIAI) was defined as localized or generalized peritonitis secondary to stomach, bowel, or biliary tract perforation; intra-abdominal surgery due to infection of another intra-abdominal organ such as the liver, spleen, pancreas, or kidneys.

High-risk peritonitis was defined as peritonitis requiring broad spectrum antibiotic coverage if the 2010 IDSA guidelines were applied. Severe CA was defined as peritonitis acquired in the community associated with APACHE II score >15, SOFA >5 or the presence of an immunocompromised state, defined as the presence of an ongoing immunosuppressive treatment, radiotherapy, chemotherapy, or high-dose steroids > 14 days, and/or a history of leukemia, lymphoma, or AIDS.

Severe HA infection involves patients with intra-abdominal severe infections developed after >48 h of hospital admission and the same level of severity scores or comorbidity.

The severity of underlying disease was measured using the comorbidity index and score of Charlson et al[7,8]. Severity of peritonitis was assessed by the APACHE II, SOFA score and the Mannheim peritonitis index[9,10]. Septic shock was defined as sepsis-induced hypotension despite adequate fluid resuscitation, with hypoperfusion or organ dysfunction requiring vasopressor drugs.

Empirical antimicrobial therapy was defined as treatment given within 24 h after diagnosis and/or surgery. It was considered adequate if active against all the cultured bacteria and/or moulds, administered intravenously at the correct doses, and given for at least 7 days, or replaced by an adequate antimicrobial within this same period.

Outcome was determined at 30 days after discharge from ICU.

Patients were considered clinically cured if all infection-related symptoms and signs had disappeared, without any evidence of a complication. Infection-related ICU mortality was defined as death during the cIAI episode in the ICU, without any other obvious cause.

Microbiology:

Peritoneal fluid was collected in sterile pots or syringes. In the absence of fluid, a smear of the peritoneal cavity was performed using an appropriate swab (AMIES, Oxoid Transport System, UK).

Samples were inoculated onto an aerobic medium (lamb blood and chocolate agar) and incubated at 35-36°C (and in 5% CO_2 for chocolate agar plates) for 48 h. Fungi were cultured on Sabouraud–chloramphenicol medium, incubated at the same temperature for 15 days. Microorganisms were mainly identified using the Vitek 2 System (bioMérieux, Marcy l'Etoile, France). Susceptibility testing was performed by automatic microdilution in plates by Microscan (Siemens), and interpretation was done according to the Clinical and Laboratory Standards Institute (CLSI) 2005 guidelines.

Analysis:

Continuous variables were represented as mean and standard deviation, tested for normal distribution with Shapiro-Wilk test and compared using Student's t-test or Mann–Whitney test, as appropriate. Categorical variables were expressed as percentages and 95% confidence intervals (95% CIs) and compared using the Chi-square test or Fisher's exact test, as appropriate. A double-sided p-value of < 0.05 was considered statistically significant.

Receiver operating characteristic (ROC) curves were used to obtain cut-off values of mortality-predicting APACHE II, SOFA scores and Mannheim index.

Results:

Analysis of all samples of peritoneal fluid cultures during the study yielded 57 episodes of cIAI.

Thirty six (63%) were severe HA peritonitis (group A), and 21 (37%) were severe CA peritonitis, group B. There were only 6 patients (17%) immunocompromised in the group A versus (vs) 1 (5%) in group B.

Mean age was 57.9 years (SD 14.6). Thirty-seven were men (65%). The mean Charlson index was 2,8 (SD 1.9). Pancreas and hepatobiliary tract were the most common origin of peritonitis (40%), followed by stomach or small bowel (28%), colon (18%), other or unknown origin (14%).

The mean Mannheim peritonitis index was 24 (SD 6.2). Mean APACHE II score was 18 (SD 8.0) and SOFA score was 7.6 (SD 4.1). The mean ICU stay was 16 days (SD 15.2) and mean days on VM was 12 (SD 11.6).

The mean time frame between the onset of infection and surgery was 2.5 days (SD 1,5). There was a markedly delay in group A (77 h vs 46 h, p = 0,002). However there was no difference related to survival in time until surgery (61 h in survivors vs 68 h in non survivors, p = 0,29).

Antibiotics against anaerobic bacteria, such as clavulanic acid, piperacillin/tazobactam, metronidazole, or meropenem, were prescribed in the majority of patients (48 patients, 84.2%). The treatment was considered adequate in 18 patients (31.5%). In 15 patients (26.3%) the isolated pathogen(s) was resistant to the previous administrated antimicrobial.

Thirty percent (17/57) needed a second surgical revision after initial surgery or new drainage because of clinical failure, but at day 30, 42% of the patients were considered cured. The mortality rate in the ICU was 40 % (23/57). The global 30-day mortality rate after ICU discharge was 5%.

Fifty five microorganisms were cultured from peritoneal samples.

Gram-negative rods were the predominant pathogens (28/55) (51%), followed by Gram-positive cocci (24%) (13/55), and moulds (13/55) (24%). Detailed microbiological results are exposed in Table 1.

Five patients (15.5%) (5/33) had associated bacteremia (all of them were Gram negative-rods, two *A. baumanii* and three *P. aeruginosa*).

Clinical, treatment, and microbiological characteristics of groups A and B are also outlined in Table 1.

Comparing the two groups (A and B) yielded no differences in age (p = 0.27), Charlson comorbidity index (p = 0.16), Mannheim peritonitis index (p = 0.48), Apache II (p = 0.69), SOFA (p = 0.27), length of ICU stay (p = 0.57), days on MV (p = 0.75), incidence of shock (p = 0,36) and mortality in ICU (p = 0.50).

There was a significant difference between two groups in relation with adequate treatment, HA (group A) 41% vs CA (group B) 82% (p = 0.034).

ROC curves were calculated to predict mortality, yielding a cut-off value of 17.5 for Apache II, 95% CI 0.682-0.559; SOFA showed a cut-off point of 7.5 95% CI 0.682-0.618; Mannheim cut-off point was 25.5 CI 95% 0.727-0.571.

Table 2 summarizes the univariate analysis for predictors of death. Patients who died had higher but not significant Charlson index (p = 0.08), similar Mannheim peritonitis index (p = 0.13), same APACHE II score (p = 0.12) and higher SOFA (p = 0.02). The survival rate was not influenced by age, or onset of infection, anatomic origin of peritonitis, being on antimicrobial therapy at the time of admission, inadequate treatment, delayed surgery, need for re-intervention, associated bacteremia, or the presence *P. aeruginosa*.

Discussion:

Our study reports 57 episodes of severe complicated intra-abdominal infections admitted to our ICU. All patients underwent different abdominal surgical procedures in order to control the source of infection.

Empirical antibiotic treatment was given to all patients several hours prior surgery. Since all patients had very high risk cIAI with high mean Apache II, SOFA and Mannheim, not all empiric antibiotic treatment followed the last 2010 IDSA guidelines.

Our results suggest that severe peritonitis have the same clinical course independently of the place of onset when it reaches a high severity score and so both of them (group A and B) go in a parallel mode. This has been found as completely different in less severe intra-abdominal infections[7,8].

Although there was a significant difference in the time to surgery in both groups, this was not related with survival. This finding was also described by Tridente et al[11] and Theunissen[12] and opposed to what it has been published before[13-17,23]. In our study this finding may be due to the fact that surgeons tend to wait longer to intervene when the patient is already on antibiotics.

It is also a fact that relaparotomy or second look surgery are very often needed. In our study nearly one third of patients went back to operating theatre due to a bad clinical course or because a definite control of the source of infection was not completely reached.

One of the aims of our study was to focus on the microbiological etiology of severe peritonitis in order to validate (in a retrospective way) the IDSA recommendation to treat severe CA or HA disease with broad-spectrum antibiotics such as meropenem, cefepime, piperacillin–tazobactam, etc. Our data on HA disease and even in CA, suggest a high Gram-negative resistance rate to narrow-spectrum antimicrobials such as amoxicillin–clavulanic acid or cefuroxime (42%), thus suggesting that healthcare and hospital risk factors of Klevens et al[18] do apply to the microbiological etiology of community-onset infections. Severe CA infections were not enough to make a conclusion about the microbiological etiology; however the 35% resistance rate of Gram-negative organisms to amoxicillin–clavulanic acid or cefuroxime may possibly justify broad-spectrum coverage.

We also have to highlight the important role that *Enterococcus* and yeasts play in the involved flora of severe peritonitis, which may justify their empirical coverage in cIAI.

Very surprising is the fact of finding more failures in correct empirical antibiotic treatments in HA peritonitis group as local microbiological flora was supposed to be better known by surgeons and critical care physicians. We have to emphasize the importance of following the IDSA recommendations and bearing in mind the local flora in order to target the microbiological results.

However, our study showed no correlation between the correct antibiotic treatment and ICU mortality. This lack of relationship between appropriate treatment and outcome might be due to insufficient study power, since several authors have previously demonstrated a higher rate of complications and/or treatment failure in cases of inadequately treated cIAI, in community, as well as in HA disease[19-21]. Not all of these studies found differences in mortality[12,19,22]. Possible reasons are the virulence of the microorganisms and their inoculum size, as well as other factors such as host defense and inadequacy of surgical management in source control, explaining failures of antimicrobial therapy even in the presence of drug susceptible organisms[4,19,22,24].

Evidence suggests that antibiotics are most effective during the early phase of infection and not later on. The incidence of septic shock was similar in CA and HA cIAI.

The similar incidence of shock in both groups support the idea that clinical course seems to depend mostly on severity of peritonitis, measured by the degree of peritoneal involvement, Mannheim index, APACHE II, and SOFA scores but not on the place of illness onset.

An Apache II, SOFA and Mannheim index higher than 17.5, 6.5 and 25 respectively (with a sensibility of 68% and specificity of 56%; 73% and 59%; 73% and 57% respectively) can help to predict mortality.

Our study, despite the limitations, supports that IDSA guidelines can be applied to cIAI in an European tertiary Hospital, particularly in severe cases candidates to be admitted in an ICU. It must also be emphasize the importance of infection source-control surgery, adequate antibiotic treatment and early resuscitation measures.

Conflict of interest: No conflict of interest to declare.

Table 1. *Clinical, treatment and microbiological characteristics of patients with complicated intra-abdominal infection classified by mode of acquisition*

Characteristic	Group A (n=36) HA	Group B (n=21) CA
M/F ratio	24 / 12	13 / 8
Age, years, mean (SD)	54,9 (16,6)	59,6 (13,3)
Mannheim Peritonitis index, mean (SD)	24,9 (6,2)	23,7(6,2)
Charlson comorbidity index, mean (SD)	3,1 (1,8)	2,3 (2)
Shock, n (%)	36 (100%)	20 (95%)
Apache II, mean	18,6	17,7
SOFA, mean	8,3	6,9
Length of ICU stay, mean (SD)	15 (15,1)	17,4 (15,6)
Death in ICU, n (%)	15 (41%)	7 (33%)
Adequate Treatment, n *(total positive samples)*	9 (22)	9 (11)
Antimicrobial Treatment[a]	n=36	n=21
Narrow-spectrum, n (%)	7	4
Broad-spectrum, n (%)	28	17
Fluoroquinolone, n (%)	1	0
Gram-positive bacteria, n	8	5
Enterococcus spp, n (%)	6 (75%)	5 (100%)
Other, n (%)	2 (25%)	0
Gram-negative bacteria, n	20	8
Enterobacteriaceae		
Escherichia coli, n (%)	6 (30%)	4 (50%)
Klebsiella spp., n (%)	4 (20%)	0
Other, n (%)	9 (45%)	2 (25%)
Pseudomona Aeruginosa, n (%)	1 (5%)	2 (25%)
Candida spp., n	11	3

M, male; F, female; SD: Standard Deviation; ICU, Intensive Care Unit.
Group A (HA): hospital complicated intra-abdominal infection.
Group B (CA): community complicated intra-abdominal infection.
[a]Narrow-spectrum: amoxicillin-clavulanic acid or cefuroxime; Broad-spectrum: third or fourth generation cephalosporin, piperacillin-tazobactam or meropenem.

Table 2. *Comparison of clinical, treatment and microbiological characteristics between non-survivors and survivors.*

Characteristic	Non-survivors (n=22)	Survivors (n=35)
Age, years, mean (SD)	61 (9,98)	57,6 (17)
Male sex, n(%)	18 (82%)	19 (54%)
Charlson Comorbidity Index, mean (SD)	3,68 (1,8)	2,34 (1,8)
Mannheim Peritonitis Index, mean (SD)	25,86 (6,4)	23,29 (5,9)
Adquisition, n (%)		
Community-onset	7 (32%)	14 (40%)
Hospital-onset	15 (68%)	21 (60%)
Shock, n (%)	22 (100%)	34 (97%)
Apache II, mean (SD)	20,68 (9)	17,0 (22)
SOFA, mean (SD)	9,23 (4,2)	6,64 (3,8)
Anatomic origin, n (%)		
Colon	6 (27%)	12 (34%)
Small bowel	4 (18%)	7 (20%)
Stomach	2 (9%)	2 (6%)
Biliary	10 (45%)	14 (40%)
ICU stay > 24h, n (%)	21 (95%)	35 (100%)
Length of ICU stay, mean(SD)	17,27 (17,7)	15,11 (13,6)
Adequate Treatment, n (%)	6 (27,2%)	12 (34,2%)
Delayed surgery (>24 h), n	16	28
Need for reintervention, n (%)	7 (32%)	9 (26%)
Enterococci, n (%)	6 (27,2%)	5 (14%)
Pseudomona Aeruginosa, n (%)	2 (9%)	1 (3%)

ICU, Intensive Care Unit

References:

1. Moss M. Epidemiology of sepsis: race, sex, and chronic alcohol abuse. Clin Infect Dis 2005;41(Suppl 7):S490 -7.
2. Intra-abdominal Sepsis: Newer Interventional and Antimicrobial Therapies Solomkin JS, Mazuski, J. Infect Dis Clin N Am 23 (2009) 593–608.
3. Cristou NV, Barie PS, Dellinger EP, Waymack JP, Stone HH. Surgical Infection Society Intra-abdominal Infection Study. Prospective evaluation of management techniques and outcome. Arch Surg 1993;128:193–9.3. Blot S, De Waele JJ. Critical issues in the clinical management of complicated intra-abdominal infections. Drugs 2005;65:1611–20.
4. Dougherty SH, Saltzein EC, Peacock JB, Mercer LC, Cano P. Perforated or gangrenous appendicitis treated with aminoglycosides: how do bacterial cultures influence management. Arch Surg 1989;124:1280–3.
5. Montravers P, Chalfine A, Gauzir R, Lepape A, Pierre Marmuse J, Vouillot C, et al. Clinical and therapeutic features of nonpostoperative nosocomial intra-abdominal infections. Ann Surg 2004;239:409–16.
6. Solomkin JS, Mazuski JE, Bradley JS, Rodvold KA, Goldstein EJC, Baron EJ, et al. Diagnosis and management of complicated intra-abdominal infection in adults and children: guidelines by the Surgical Infection Society and the Infectious Diseases Society of America. Clin Infect Dis 2010;50:133–64.
7. Charlson ME, Pompei P, Ales KL, MacKenzie CR. A new method of classifying prognostic comorbidity in longitudinal studies: development and validation. J Chron Dis 1987;40:373–83.
8. Charlson M, Szatrowski TP, Peterson J, Gold J. Validation of a combined comorbidity index. J Clin Epidemiol 1994;47:1245–51.
9. Linder M, Wacha H, Feldmann U, Wesch G, Streifensand FA, Gundlach E. The Mannheim peritonitis index. An instrument for the intraoperative prognosis of peritonitis. Chirurg 1987;58:84–92.
10. Notash AY, Salimi J, Rahimian H, Fesharaki MH, Abbasi A. Evaluation of Mannheim peritonitis index and multiple organ failure score in patients with peritonitis. Indian J Gastroenterol 2005;24:197–200.
11. Tridente A et al. Patients with faecal peritonitis admitted to European intensive care unit: an epidemiological Surrey of the GenOSept cohorte. Intensive Care Med 2014;40:202-210.
12. Theunissen C, Cherifi S, Karmali R. Management and outcome of high-risk peritonitis: a retrospective survey 2005–2009. International Journal of Infectious Diseases 15 (2011) e769–e773.
13. Shlaes DM, Gerding DN, John JF Jr, et al. Society for Healthcare Epidemiology of America and Infectious Diseases Society of America Joint Committee on the Prevention of Antimicrobial Resistance: guidelines for the prevention of antimicrobial resistance in hospitals. Clin Infect Dis 1997;25:584–99.

14. Koperna T, Schulz F. Prognosis and treatment of peritonitis. Do we need new scoring systems? Arch Surg 1996;131:180–6.
15. Koperna T, Schulz F. Relaparotomy in peritonitis: prognosis and treatment of patients with persisting intraabdominal infection. World J Surg 2000;24:32–7.
16. Mulier S, Penninckx F, Verwaest C, et al. Factors affecting mortality in generalized postoperative peritonitis: multivariate analysis in 96 patients. World J Surg 2003;27:379–84.
17. Grunau G, Heemken R, Hau T. Predictors of outcome in patients with postoperative intra-abdominal infection. Eur J Surg 1996;162:619–25.
18. Klevens RM, Morrison MA, Nadel J, Petit S, Gershman K, Ray S, et al. Invasive methicillin-resistant Staphylococcus aureus infections in the United States. JAMA 2007;298:1763–71.
19. Mosdell DM, Morris DM, Volture A, Pitcher DA, Twiest MW, Milne RL, et al. Antibiotic treatment for surgical peritonitis. Ann Surg 1991;214:543–9.
20. Montravers P, Gauzit R, Muller C, Marmuse JP, Fichelle A, Desmonts JM. Emergence of antibiotic-resistant bacteria in cases of peritonitis after intraabdominal surgery affects the efficacy of empirical antimicrobial therapy. Clin Infect Dis 1996;23:486–94.
21. Falagas ME, Barefoot L, Griffith J, Ruthazar R, Snydman DR. Risk factors leading to clinical failure in the treatment of intra-abdominal or skin/soft tissue infections. Eur J Clin Microbiol Infect Dis 1996;15:913–21.
22. De Waele JJ, Hoste EA, Blot S. Blood stream infections of abdominal origin in the intensive care unit: characteristics and determinants of death. Surg Infect (Larchmt) 2008;9:171–7.
23. Azuhata et al. Time from admission to initiation of surgery for source control is a critical determinant of survival in patients with gastrointestinal perforation with associated septic shock. Critical Care 2014, 18:R87.
24. Roehrborn A, Thomas L, Potreck O, Ebener C, Ohmann C, Goretzki PE, et al. The microbiology of postoperative peritonitis. Clin Infect Dis 2001;33:1213–9.

Peptide extracts of different Sardinia cheeses might exhibit antimicrobial and antifungal activity after the *in vitro* digestion of the products

Filomena Nazzaro*, **Raffaele Coppola and Florinda Fratianni**

ISA-CNR Institute of Food Science Via Roma 64,83100 Avellino, Italy
*Corresponding author: e-mail: mena@isa.cnr.it, Phone: +390825299102

The aim of our work was to investigate the biological potential of some dairy products typical of the Sardinia region, in particular: "Saboriu" Pecorino cheese, obtained from raw milk and raw pasta with live autochthonous lactic acid bacteria; "Callau axeddu" soft cheese, made with live autochthonous lactic acid bacteria; "Gioddu", a typical natural Sardinian yogurt produced with raw sheep milk fermented with live autochthonous lactic acid bacteria. The study was focused on the capability of peptide extracts to exhibit antimicrobial and antifungal activity, after the *in vitro* gastrointestinal transit of products. The most effective antimicrobial activity was exhibited by peptide extracts of Gioddu and Saboriu, which were effective in inhibiting the growth of *Escherichia coli*, *Pseudomonas aeruginosa* and the emerging pathogen *Chromobacter sakazakii*. Callau axeddu inhibited the growth of *E. coli* and *C. sakazakii*. Peptides of Callau show antifungal activity against *Aspergillus niger* and *A.versicolor*, The extract of Saboriu inhibited *Aspergillus* and *Penicillium digitatum*.

Keywords dairy peptides; antimicrobial activity

1. Introduction

General remarks

The protection of the traditional agricultural foodstuff is a key point for the preservation and enhancement of the Italian productive and socio-cultural environments, mainly in some regions, such as Sardinia, where it represents a valuable resource for the regional economy. Unfortunately, some products, such as cheese, are now at risk of being marginalized partly because of the crisis that is affecting the Italian food industry. Sardinian dairy products are of particular interest also by a scientific viewpoint for different reasons: the island is geographically isolated; there are several small farms where sheep and cows are bred in semi-wild conditions; often, the widespread production of traditional dairy products is on a small scale. Therefore, a series of plans are necessary to safeguard and enhance the market of the traditional products, also through a study for the identification and exploitation of their functional properties, such antimicrobial activity against different pathogens. The aim of our work was to investigate the biological potential of some dairy products typical of the Sardinia region. The work was performed on three organic cheeses, the "Saboriu" Pecorino cheese, (seasoning 3 months), obtained from raw milk and raw pasta with live autochthonous lactic acid bacteria; the "Callau or Casu Axedu" soft cheese, made with live autochthonous lactic acid bacteria; and the "Gioddu", a typical natural Sardinian yogurt produced with raw sheep milk fermented with live autochthonous lactic acid bacteria (Figure 1). The technology of production for this cheese is not well defined as different methods and types of milk are used depending on local tradition. These three kinds of cheese in Sardinia have a long tradition of being regarded as a healthy product, as other fermented milks in other parts of the world [1]. New bioactive peptides with bioactivity could be discovered, which could provide greater health benefits to consumers. The study was focused on the identification of peptides with antimicrobial activity, obtained after the *in vitro* gastrointestinal transit of products, treatment with acetonitrile to remove the protein fraction, and drying of samples to remove the organic solvent. The antimicrobial activity was assessed by the inhibition halo test using *Bacillus cereus*, *Escherichia coli*, *Pseudomonas aeruginosa*, and *Chromobacter sakazakii* as pathogen strains; the antifungal activity was assessed versus *Penicillium expansum*, *Penicillium griseofulvus*, *Aspergillus niger*, *Aspergillus versicolor*, *Penicillium digitatum*, *Penicillium citrinum*.

2. Materials and methods

2.1 "Callau Axedu"soft cheese, "yogurt Gioddu" and "Saboriu" pecorino cheese (Figure 1 from left to right, respectively) were purchased by an organic plant in the province of Cagliari, Sardinia, Italy. Samples were processed simulating the gastro-intestinal passage, with a first step for 3 h at 37°C in sterile 0.85% NaCl+ pepsin, pH 2.5 [2] then for 1 h at 37°C in sterile 0.85%NaCl+ bile salts+pancreatin, pH 9.0. The supernatant, recovered by centrifugation, was treated with 3 volumes of acetonitrile [3] and kept for 24 h at 4°C, so to discard proteins. After centrifugation, supernatant, containing peptides, was recovered, and dried. Each sample was then resuspended in sterile 0.85% NaCl for the microbial test.

Figure 1   from the left to right "CALLAU AXEDU" SOFT CHEESE; "GIODDU" YOGURT; "SABORIU" PECORINO CHEESE

2.2 Antimicrobial activity

Gram-positive *Bacillus cereus* and Gram-negative *Escherichia coli*, *Pseudomonas aeruginosa* and *Chromobacter sakazakii* were the pathogenic type strains used for the antimicrobial test; they were obtained from the DSMZ. Each strain was incubated at 37 °C for 18 h in tryptone yeast extract (Oxoid, Milano, Italy). Different fungal strains of agro-food interest, the citrinin-producing strain *Penicillium citrinum*, *Penicillium griseofulvus*, *Penicillium expansum*, *Penicillium digitatum*, *Aspergillus niger*, *Aspergillus versicolor* were used to evaluate the antifungine activity. All strains were purchased from DSMZ (Deutsche Sammlung von Mikroorganismen und Zellkulturen GmbH, Braunschweig, Germany). To test the antibacterial activity, a cell suspension of bacteria (previously grown in Nutrient Broth at 37°C for 18h) was spread onto Nutrient Agar plates. Different volumes of samples (final doses ranging from 5 to 25 µl/paper disc), were put onto sterile filter disks (diam 5 mm). A cell suspension of fungi was prepared in sterile distilled water and plated onto Potato Dextrose Agar (PDA) (Oxoid). After 20 min under sterile conditions at room temperature, plates were incubated at 28°C until the mycelium of fungi reached the edges of the control plate (negative control, without the sample added extracts); the resulting clear zones of inhibition were measured in mm, expressing the antifungal activity. DMSO (10µL) was used as negative control. Samples were tested in triplicate and the results are expressed as mean ± standard deviation.

3. Results

In Figure 2 is shown the antimicrobial activity expressed by peptide fraction of Callau Axedu soft cheese (a),Gioddu yougurt (b) and Saboriu pecorino cheese (c)

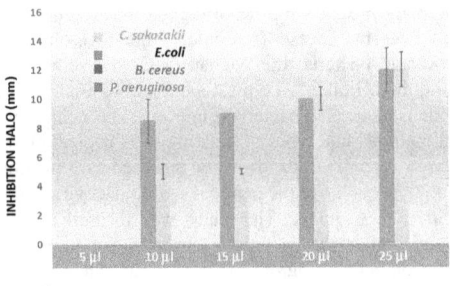

a)

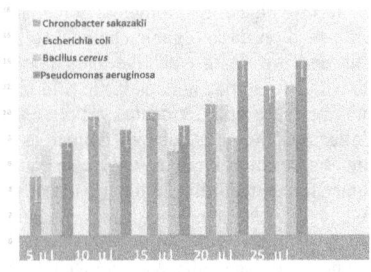

b)

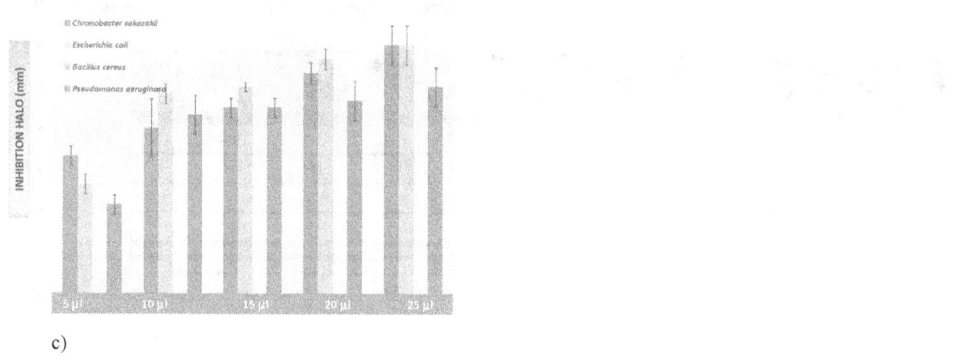

c)

Figure 2. Antimicrobial activity exhibited by peptide extracts of Callau axedu (a), Gioddu (b) and Saboriu (c). The data are reported as mm of inhibition obtained using a volume ranging from 5 to 25 µl of sample.

The three extracts showed a different behaviour against pathogens used as tester strains. Callau axedu exhibited antimicrobial activity mainly *versus* the emergent pathogen *C. sakazakii* and the enterotoxinogenic *E.coli*, with inhibition halo ranging from 5 to 12 mm. Peptides of yogurt Gioddu were active against all the four pathogens and with each volume used into the experiments. The most effective activity was shown against *P. aeruginosa*, with inhibition halos which reached 14 mm, using the highest volume of the extract. Saboriu pecorino cheese was ineffective against *B.cereus*; positive results were observed using its peptide extract *versus E. sakazakii*, *P. aeruginosa*, and *E.coli*. These last two strains were more sensitive to the presence of the extract than *C. sakazakii*, giving rise to an inhibition halo of 11 mm, in the presence of 25 µl of the extract. Such result could be considered of a certain interest, also taking into consideration the gastrointestinal transit (although simulated), a process which is not often used in the study of the characteristics and biological properties of dairy peptides. Cheeses are a big source of bioactive peptides, taking into account the very strong activity exhibited by microbial or endogenous enzymes on milk and dairy proteins. During the aging of cheese (ripening and mauturation), microbial flora (starter and non starter bacteria) can affect the cheese quality, contributing also to eventually enrich the products with bioactive peptides [4], with different activity. Some Italian cheeses, including Asiago d'Allevo [5] cheese demonstrated antimicrobial activity; on the other hand, Pecorino Toscano-like cheese and other fermented ovine cheeses did not showed antimicrobial activity [6].

3.1 Antifungal activity

Figure 3 shows the antifungal activity (expressed as mm of inhibition halo) exhibited by the extracts of Callau axedu (a), and Saboriu (b) against the pathogen fungi used as tester. In Table 1 are reported the negative data of the antifungal activity obtained testing the extract of yogurt Gioddu.

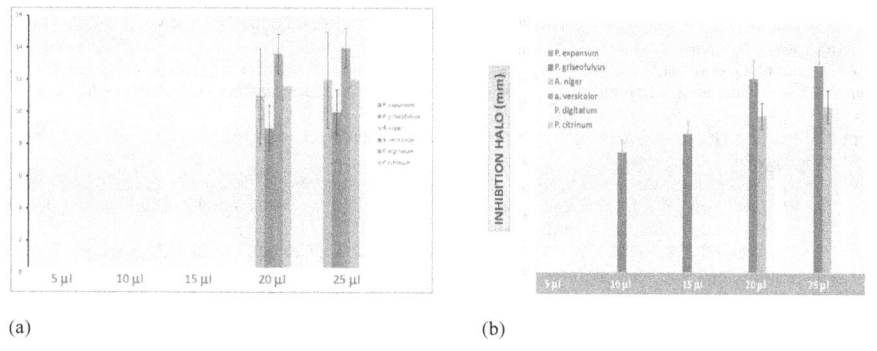

(a)   (b)

Figure 3. Antifungal activity exhibited by peptide extracts of Callau axedu (a), and Saboriu (b). The data are reported as mm of inhibition obtained using a volume ranging from 5 to 25 µl of sample.

Table 1 Evaluation of the antifungal activity exhibited by peptide extracts of yogurt Gioddu using different (from 5 to 25 µl) volumes.

Volume	*P. expansum*	*P. griseofulvus*	*A. niger*	*A. versicolor*	*P. digitatum*	*P. citrinum*
5µl	-	-	-	-	-	-
10 µl	-	-	-	-	-	-
15 µl	-	-	-	-	-	-
20 µl	-	-	-	-	-	-
25 µl	-	-	-	-	-	-

The antifungal test demonstrated a marked difference of behaviour among the three peptide extracts, conversely to that observed against the four pathogen bacteria, whichwere all sensitive to the activity of the extracts, although in different ways. In particular, the extract of yogurt Gioddu demonstrated a complete absence of activity against all 6 mould strains (Table 1). By analysing the behaviour shown by the extracts of Callau axedu (soft cheese) and Saboriu (pecorino cheese, thus with a prolonged time of aging), we could hypothesise that, in some cases, the antifungal activity increased with the aging of the product: in fact, peptides of Callau axedu demonstrated activity against *A. niger*, *A.versicolor*, *P.digitatum* and *P.citrinum*, (with the two highest volumes, 20 µl and 25 µl, used in the experiments), but not against *P. expansum* and *P. griseofulvus*, which, on the contrary, were sensitive to the presence of peptides obtained from Saboriu, and to almost all volumes used into the experimentation (except the lowest, 5 µl). Also in this case, although several papers reported the antifungal activity of peptides giving rise from different dairy products [7], at least at our knowledge, there are few reports regarding the fate and the biological properties of antifungal peptides after the gastrointestinal transit, neither in vitro studies are reported.

3.2 Conclusions

This study showed that such three dairy products, with different aging, typical of the Sardinia region, can represent an interesting source of bioactive peptides, useful not only for their antimicrobial but also for their action against unwanted and/or pathogenic moulds. Therefore, findings of the present work suggest also that dairy peptides with concurrent antimicrobial and antifungal properties could have a potential for practical applications, and could be used both for pharmaceuticals purposes and as natural preserving agents.

References

[1] Mathara JM, Schillinger U, Kutima PM, Mbugua KS, Holzapfel WH. Isolation, identification and characterization of the dominant microorganisms of kule naoto: The Maasai traditional fermented milk in Kenia. International Journal of Food Microbiology, 2004; 94: 269–278

[2] De Giulio B, Orlando P, Barba G, Coppola R, De Rosa M, Sada A, De Prisco PP, Nazzaro F. Use of alginate and cryo-protective sugars to improve the viability of lactic acid bacteria after freezing and freeze-drying. World Journal of Microbiology and Biotechnology,2005; 21: 739-746.

[3] Nazzaro F, Fratianni F, Nicolaus B, Poli A, Orlando P. The prebiotic source influences the growth, biochemical features and survival under simulated gastrointestinal conditions of the probiotic *Lactobacillus acidophilus*. Anaerobe, 2012; 18: 280–285

[4] Settanni L., Moschetti G. Non-starter lactic acid bacteria used to improve cheese quality and provide health benefits. Food Microbiology 2010; 27: 691–697.

[5] Lignitto L, Segato S, Balzan S, Cavatorta V, Oulahal N, Sforza S, Degraeve P, Galaverna G, Novelli E. Preliminary investigation on the presence of peptides inhibiting the growth of *Listeria innocua* and *Listeria monocytogenes* in Asiago d'Allevo cheese. Dairy Science & Technology,2012; 92: 297-308.

[6] Meister Meira SM, Daroit DJ, Etges Helfer V, Folmer Corrêa AP, Segalin J, Carro S, Brandelli A. Bioactive peptides in water-soluble extracts of ovine cheeses from Southern Brazil and Uruguay. Food Research International,2012; 48: 322– 329

[7] Théolier J, Hammami R, Fliss I, Jean J. Antibacterial and antifungal activity of water-soluble extracts from Mozzarella, Gouda, Swiss, and Cheddar commercial cheeses produced in Canada. Dairy Science & Technology,2014; 94: 427-438.

Retreatment of relapsing small intestinal bacterial overgrowth with Rifaximin α polymorph is effective and safe

Lombardo Lucio[1,2], Schembri Mariangela[3]

1. Gastroenterology Service, Poliambulatorio Statuto, Piazza Statuto 3 ang. Via Manzoni 0, 10100-Turin (Italy)
2. Gastroenterology Dpt, Mauriziano U.1st Hospital, Corso Turati 42, 10128-Turin (Italy), from 1977 to October 2012.
3. Gastroenterology Dpt, Mauriziano U 1st, Corso Turati 42, 10128-Turin (Italy)

Background: The incidence of small intestinal bacterial overgrowth (SIBO) is increasing, mainly because of the increase of pharmacological risk factors. Although SIBO can be successfully eradicated by Rifaximin, its recurrence is easily predictable, as long as the risk factors persist, as long-term therapy with proton pump inhibitors (IPP), chronic atrophic gastritis, lactose intolerance etc. Re-treatment therefore can be a clinical challenge. *Material & Method*: One hundred and forty four patients treated on long-term treatment with PPI for gastro-esophageal reflux disease (GERD), successfully eradicated from SIBO with high dose Rifaximin were followed-up for 1 year for relapse investigation. At the end of follow-up, or before if symptoms suggested it, glucose hydrogen breath test (GHBT), (Quintron, Milwaukee, WI, USA) was performed to each patient and symptoms were recorded. All relapsed patients were retreated with Rifaximin 1200 mg/die for 2 weeks, as in the first course. The outcome of therapy was assessed both clinically and by means of GHBT, 2 months after the completion of Rifaximin course. *Results*: Out of a cohort of 144 patients (M 84; mean age 46±14) successfully eradicated from SIBO with Rifaximin, 52 patients (M 32; mean age 45±13) relapsed (36%), mainly because of continuation of treatment with PPI for GERD (49/97, i.e. 50.5%). Forty-seven out of 52 patients retreated with Rifaximin showed negative GHBT along-side symptoms remission (90%), indicating a successful eradication from SIBO. No relevant side effect was registered. *Conclusions*: 1) Relapse rate of SIBO within 1 year is high if treatment with PPI is not discontinued. 2) Retreatment with Rifaximin 1200 mg/die for 2 weeks results to be effective and safe.

Keywords: SIBO; PPI; GHBT; Rifaximin; retreatment

1. Introduction

The incidence of small intestinal bacterial overgrowth (SIBO) has lately been shown to be exponentially increasing, mainly because of the increase of pharmacological risk factors [1, 2].

Although SIBO can be successfully eradicated by Rifaximin in a high percent of the patients [1], its recurrence is easily predictable as long as risk factors persist as long-term therapy with proton pump inhibitors (PPIs), chronic atrophic gastritis, lactose intolerance [3] etc. Retreatment therefore can be a clinical challenge. The aims of the study are: 1) Evaluate the relapse rate of SIBO. 2) Assess the re-eradication rate by Rifaximin.

2. Material & Methods

2.1 SIBO Diagnosis

Direct (duodenal-jejunal aspirate culture) and indirect tests (hydrogen breath tests) are useful tools for SIBO diagnosis with relative advantages and disadvantages. We used glucose hydrogen breath test (GHBT; Quintron Milwaukee, WI, USA), because it is noninvasive, reliable, repeatable and easy to perform, [1]. In a previous study on 450 subjects, we evaluated the increased risk of SIBO during long-term treatment with PPIs, by means of GHBT. Then we treated SIBO-positive patients with Rifaximin 1200 mg a day for two weeks, with an eradication rate of 90% [1]. Confirmation of increased risk of SIBO by PPIs came from a meta-analysis study published in 2013 [4].

2.2 Patients

We then performed a follow-up surveillance study for SIBO relapse on one hundred and forty four patients, previously treated with PPIs for Gastro-Oesophageal Reflux Disease (GORD), affected by SIBO and successfully eradicated with 2 weeks high dose Rifaximin α polymorph. Ninety-seven patients further continued PPI treatment and 47 discontinued it. Twenty additional patients with IBS and no PPI treatment, as well affected and eradicated from SIBO, entered the study. At the end of follow-up, or before if symptoms suggested the occurrence of SIBO (like diarrhea, abdominal pain, bloating), each subject was submitted to GHBT.

2.3 Statistics

Statistical analysis of data was carried out by SPSS software. For quantitative variables the Mann-Whitney test was used. The X^2 test with Yates correction was performed to evaluate SIBO prevalence.

2.4. Retreatment

All the relapsed patients were retreated with Rifaximin α polymorph 1200 mg a day for 2 weeks, as in the first course.

3. Results

Out of a cohort of 144 patients previously treated with long-term PPIs for GORD (M 89, F 55; mean age 46±14 years) and successfully eradicated from PPI-related SIBO with high dose Rifaximin therapy, 97 patients continued and 47 discontinued therapy with PPIs. As a whole, 52 patients (M35, F17; mean age 45±13 years) relapsed (36%), at a mean time of 8±4 months after the first successful eradication. Subdividing by subgroups, we registered a SIBO relapse in 49 out of 97 patients continuing PPI treatment, in 3 out of 47 patients discontinuing it and in 1 out of 20 IBS patients within the year of the observation (50.5%, 6% and 5%, respectively) with a statistically significant difference between the first group and the others, p<.001). Further analyzing the groups of patients on PPI treatment according to the duration of treatment with PPIs before the first eradication of SIBO, the relapse rate resulted higher in patients with > 1 year of treatment (*25% among patients with <1 year of treatment and, respectively, 57%, 59% and 50 % among patients who previously assumed PPIs for >1<3 , >3<5 and >5-10 years*). The differences between the first and the other groups were statistically significant (p<.005). Rifaximin is a poorly absorbed antibiotic with a broad spectrum of bactericidal activity, well known around the world.

Forty-seven out of the 52 SIBO relapsed patients, retreated with high dose Rifaximin, showed negative GHBT along-side symptoms remission, indicating successful SIBO eradication (90%), with the same successful rate as in the first course of treatment. No relevant side effect was registered.

4. Discussion

The results of our study show that: 1) Relapse of PPI-related SIBO within 1 year is frequent if treatment with PPI is not discontinued (50.5% of patients). Our data are in keeping with those of Lauritano et al., described in patients affected by IBS as predisposing condition [5] 2) The relapse rate is higher in patients with previous PPI treatment lasting for more than 1 year. 3) Retreatment with Rifaximin α polymorph at dosage of 1200 mg a day for 2 weeks results to be effective and safe. SIBO is an emerging clinical problem as long-term treatment with PPIs is very common in everyday medical practice, also for inappropriate over the counter medication [6]. In this context, the results of the present study are of practical usefulness in that can assist the clinician in two ways: 1) first of all, by alerting the clinician to avoid long-term treatment with PPIs, especially if inappropriate, encouraging the use of alternative, "local", non-systemic therapy for GORD. Alternatively, seasonal or *à la démande* PPI treatment could be worth of trying. 2) Secondly, relapse of SIBO, if occurs, can be confidently treated by Rifaximin α polymorph, that has not shown clinical antibiotic resistance nor selection of resistant bacterial strains nor relevant systemic side effects, in this context, till now. As a matter of facts, Rifaximin seems to be able to modulate the gut microbiota rather than cause bacterial resistance [7]. In a recent meta-analysis on antibiotic therapy for SIBO, Rifaximin is reported as the most commonly studied antibiotic with overall breath test normalization rate of 49.5% [8]. Our data are quite divergent from this value, probably because of differences in the regimen and dose of Rifaximin, and because of the possibility of publication selection bias. In this respect, we outline that, in the cited meta-analysis, prospective open studies with pre-test/post-test design, as our own one, are not included. Therefore, in our opinion, it is worth suggesting that the dosage of 1200 mg/day and the duration of two weeks of treatment should be assumed as the standard of cure for SIBO treatment, under these conditions, as a consequence of the best-obtained results. In conclusion, we think that Rifaximin 1200 mg/day for 2 weeks is effective and safe both in first and in the recurrence course of SIBO therapy. However, awareness should be stressed on discontinuing pharmacological risk factors for SIBO.

References

[1] Lombardo L, Foti M, Ruggia O, Chiecchio A. Increased incidence of small intestinal bacterial overgrowth during proton pump inhibitor therapy. *Clinical Gastroenterology & Hepatology*. 2010;8:504-8.

[2] Pilotto A, Franceschetti M, Vitale D et al. for FIRI and SOFIA project. The prevalence of diarrhea and its association with drug use in elderly outpatients: a multicentre study. *American Journal Gastroenterology*. 2008;103:2816-23.

[3] Lombardo L, Giostra A, Schembri M. High prevalence of small intestinal bacterial overgrowth in lactose intolerance patients: is it a chicken and eg situation? *British Journal of Medicine & Medical Research*. 2014;4(15):2931-9.

[4] Lo W-K, Chan W W. Proton pump inhibitor use and risk of small intestinal bacterial overgrowth: a meta-analysis. *Clinical Gastroenterology & Hepatology*. 2013;11(5):483-90.

[5] Lauritano E C, Gabrielli M, Scarpellini E, Lupascu A et al. A small intestinal bacterial overgrowth recurrence after antibiotic therapy. *American Journal Gastroenterology*. 2008;103:2031-5.

[6] Lombardo L. L'uso degli inibitori della pompa protonica a Torino. I risultati di uno studio all'ospedale Mauriziano. *Torino Medica*. 2012;10:40-2.

[7] Maccaferri S, Vitali B, Klinder A, Kolida S et al. Rifaximin modulates the colonic microbiota of patients with Crohn's disease: an in vitro approach using continuous culture colonic model system. *Journal Antimicrobial Chemotherapy*. 2010;65(12):2556-65.

[8] Sha S C, Day L W, Somsouk M, Sewell L. Meta-analysis: antibiotic therapy for small intestinal bacterial overgrowth. *Alimentary Pharmacology and Therapeutics*. 2013;38:925-34.

Role of *Candida sp.* in inflammatory respiratory diseases

A. P. Godovalov[1,2*] and L. P. Bykova[1]

[1]Department of microbiology and virology, Department of immunology, Acad. E.A. Wagner Perm State Medical Academy, 85 Ekaterininskaya str., 614990 Perm, Russian Federation

[2]The Medical Unit of the Internal Affairs Directorate in Perm region, 128 Permskaya str., 614990 Perm, Russian Federation

*Corresponding author: e-mail: AGodovalov@gmail.com, Phone: +7 9129815100

The data about the participation of *Candida* in inflammatory respiratory diseases is sometimes varied, and information on the microbial associations is insufficient. The aim of our study was to evaluate the role of *Candida* in inflammatory diseases of respiratory tract and its sensitivity to antimycotic agents. Studies have shown a high rate of detection of *Candida* in sputum. The most stable strains of *Candida* were isolated from associations with streptococci. Streptococci isolated from sputum in association with *Candida*, showed high resistance to antibiotics. In the presence of resistant *Candida* strains in the discharge of the upper respiratory tract, streptococci typically showed sensitivity to most antibiotics.

Keywords *Candida*; fungi; sputum; microbial associations; sensitivity to antibiotics

1. Introduction

Currently, we can see an extension of potentially pathogenic yeast-like fungi *Candida* in patients and healthy individuals, as well as an increased incidence of candidiasis, which takes more than 15% in the structure of infectious and inflammatory diseases [1, 5]. Such a phenomenon is associated with a wide and often irrational use of antibiotics, as well as the confounding influence of environmental factors on the resistance and the human immune system [6, 7]. This creates optimal conditions for the development of deep mycosis [5], which causative agents are opportunistic fungi previously considered nonpathogenic and widespread in the environment. The participation of *Candida* in inflammatory diseases of the respiratory tract, where they can get from the throat, mouth, together with other microorganisms is of particular interest [4, 6]. The specific gravity of pharyngomycosis is 15% [2]. It is shown that *Candida* are often found in the sputum in case of different lung inflammatory diseases (from 2 to 70% of cases) [3]. However, the information about the participation of *Candida* in inflammatory respiratory diseases is sometimes varied, and information on the microbial associations is insufficient [5]. The study of sensitivity of *Candida* in inflammatory diseases of respiratory tract to antimycotic agents is of interest.

In this regard, the aim of our study was to evaluate the effect of *Candida* in inflammatory diseases of respiratory tract and sensitivity to antimycotic agents. To solve this purpose we conducted a descriptive study of the etiological role of *Candida* in the development of inflammatory diseases of the respiratory tract, as well as microbial associations in this compartment.

2. Materials and methods

We studied 277 samples of sputum of patients with community-acquired pneumonia, as well as 389 samples of discharge of patients with inflammatory diseases of the upper respiratory tract.

Isolation of Candida from clinical specimens was performed with bacteriological method using Sabouraud's medium. Identification was performed by culture and biochemical characteristics. Blood and yolk-salt agars, Endo and thioglycolic mediums were used to isolate the accompanying bacterial flora. Determination of sensitivity of the isolated microorganisms was performed with disk-diffusion method.

3. Results

During the studies, it was found that *Candida* was found in 35.0% of sputum samples, and in 11.5% of samples – in diagnostically significant amount (10^5 colony forming units per milliliter, hereinafter referred to as CFU/mL). During the trial of the discharge of the upper respiratory tract *Candida* were found in 15.2% of samples, and in 5.6% of samples – in 10^5 CFU/ml or above. With more frequent isolation of *Candida* from the sputum, in monoculture they were more frequent in the discharge of the upper respiratory tract (27.3% of samples) than in the sputum (9.4% of samples). Distribution of *Candida* in sputum was as follows: *C. albicans* - 75% of cases, *C. krusei* – 12.5%, *C. tropicalis* – 6.2%, *C. pseudotropicalis* – 6.2%. 59.1% of cases -

C. albicans, 18.2% - *C. pseudotropicalis*, 13.6% - *C. krusei*, in 4.5% - *C. quilliermondii* were isolated from discharge of the upper respiratory tract.

In the sputum, as well as in the discharge of the upper respiratory tract *Candida* more frequently occur in association with other microorganisms. Associations with one bacterial species were identified in 52.4 and 43.7% of samples, respectively. In associations, bacteria were detected at 10^5 CFU/ml and above (90.9% and 100% of samples respectively). 43% of sputum samples contained the associations of *Candida* with two bacterial species. Among them, 22.2% of samples had one species in diagnostically significant amount, 55.5% - both bacterial species were in the diagnostic quantities, and 22.2% of samples had no bacterial species in diagnostically significant amount. Such associations in the discharge of the upper respiratory tract were found in 37% of samples. Herewith, 16.7% of samples had one bacterial species at 10^5 or more CFU/ml. 83.3% of samples had two bacterial species, each at 10^5 or more CFU/ml. *Candida* species in association with three bacterial species more frequently occur in inflammatory processes in the upper respiratory tract (12.5% of samples) than in the lower one (4.8% of samples). In these associations all bacterial species were in amount of 10^5 CFU/ml and more.

In associations of *Candida* with bacteria in inflammatory diseases of the respiratory tract, gram-positive cocci dominated (80.9% of sputum samples and 85.7% of samples of the upper respiratory tract discharge). All strains of gram-positive cocci isolated from association with *Candida* were present in the pathological material in diagnostically significant amount. Among the gram-positive cocci in association with *Candida*, streptococci dominated (78.6% of sputum samples and 83.3% of samples of the upper respiratory tract discharge). Staphylococci in association with *Candida* were present in the sputum (42.9% of samples) more often than in the discharge of the upper respiratory tract (16.7%). In sputum samples, associations of *Candida* with staphylococci and streptococci simultaneously occurred more frequently.

During the analysis of sputum samples, the associations of *Candida* and gram-negative bacteria were detected in 33.3% of cases, and during the analysis of the upper respiratory tract discharge - in 14.3% of cases. The spectrum of gram-negative bacteria was most extensive in the sputum, where representatives of *Escherichia* (20% of samples), *Klebsiella* (20% of samples), *Enterobacter* (20% of samples), and non-fermentative gram-negative bacteria (40% of samples) were detected. In these associations, gram-negative bacteria in the diagnostically significant quantity were found in 71.4% of samples. In the discharge of the upper respiratory tract, only *Klebsiella* representatives were found among gram-negative bacteria.

Study of the sensitivity of fungi to antifungal agents has shown that *Candida* isolated from sputum in association with staphylococci were sensitive to two antifungal agents in 66.7% of cases, and to one antifungal agent in 33.3% of cases. *Candida* isolated from the association with streptococci were sensitive to three agents in 12.5% of cases, to two agents - in 37.5% of cases, and to one agent - in 25% of cases. In 25% of cases, there was the resistance of *Candida* to three agents. In associations with staphylococci and streptococci, *Candida* species in all cases were sensitive only to two agents. In 50% of cases of associations with non-fermentative bacteria, *Candida* were sensitive to two antifungal agents and in 50% of cases – to one agent. In association with the representatives of *Enterobacteriacea*, *Candida* were sensitive only to two agents.

Candida, isolated from the discharge of the upper respiratory tract in association with staphylococci were sensitive to three antifungal agents in all cases. In isolation from the association with streptococci *Candida* were sensitive to three antifungal agents in 50% of cases, in 25% of cases they were sensitive to two antifungal agents and in 25% of cases - sensitive to only one antifungal agent. In associations with staphylococci and streptococci *Candida* were sensitive only to one agent in all cases. All *Candida*, isolated in association with *Klebsiella* were sensitive to three antifungal agents.

As during the study of sputum samples the most stable (25% of the isolates were resistant to three agents) and most sensitive (*Candida* were sensitive to three agents only in association with streptococci) strains of *Candida* were isolated from associations with streptococci, it was interesting to study the sensitivity of streptococci to antibiotics. Streptococci isolated from sputum in association with *Candida*, sensitive to three antimycotics, detected at 10^5 CFU/ml and more, and wherein all showed resistance to two agents. Streptococci isolated in association with *Candida*, sensitive to two agents, only in 50% of the samples were detected in a diagnostically significant amount of 10^5 CFU/ml and more. In these associations, streptococci were resistant to three agents in 33.3% of samples, - to five drugs in 33,3% of samples and - to seven agents in 33.3% of samples. In association with *Candida*, sensitive to one agent, streptococci were found at 10^5 CFU/ml and more, and a half of these strains was resistant to three agents and another one was resistant to six agents. In associations with *Candida*, resistant to three agents, streptococci were found in amount of 10^5 CFU/ml and more, half of them was resistant to two agents, and another half - to six agents.

When studying the properties of microorganisms isolated from the discharge of the upper respiratory tract, it was found that all of streptococci from association with *Candida*, sensitive to three drugs were detected at 10^5 CFU/ml or more. Half of streptococci strains showed resistance to four agents, and the other half was resistant to two agents. In 50% association with *Candida*, sensitive to two drugs, streptococcus were found at 10^5 CFU/ml or more. Those strains of streptococci which quantity was 10^5 CFU/ml or more, in all cases, were resistant to three agents. In associations represented by *Candida*, resistant to three agents, streptococci were found in amount of 10^6 CFU/ml or more, and they all showed the resistance to two agents.

Studies have shown a high frequency of detection of *Candida* in sputum in inflammatory diseases of the respiratory tract. *C. albicans* were the most common. The inflammatory process was supported by bacterial flora, with a predominance of potentially pathogenic species. Fungal-microbial associations were detected in most cases. Combinations of fungi with streptococci and staphylococci were registered in most associations. Massiveness of content of biological material was high. We revealed the resistance of fungi to antifungal agents, which have varied in microbial associations. *Candida* were more resistant in associations with *Streptococcus*. Streptococci isolated from the sputum with resistant strains of *Candida* also showed high resistance to antibiotics. During the analysis of the results of antimicrobial sensitivity of microorganisms strains isolated from the discharge of upper respiratory tract, the situation was different: in the presence of resistant strains of *Candida*, streptococci generally showed sensitivity to most antibiotics.

4. Conclusion

Thus, these data suggest the participation of fungal-bacterial associations in inflammatory diseases of the respiratory tract. In the discharge of the upper respiratory tract and sputum it was revealed not a pure culture of the pathogen, but the combination of different microbes, most of *Candida* with *Streptococus* and *Staphylococcus*. Participants of fungal-streptococcal associations exhibited high resistance to antimicrobial agents. Revealed facts may indicate the possibility of mutual influence of microbes-associates and exchange of genetic information between them. It's necessary to take into account the possible role of each of associates in the pathological process to ensure the effective treatment of inflammatory diseases of the respiratory tract.

References

[1] Bassetti M, Righi E, Ansaldi F. et al. A multicenter study of septic shock due to candidemia: outcomes and predictors of mortality. Intensive Care Med. 2014; 40:839-845.
[2] Kunel'skaia V, Shadrin GB. The modern approach to diagnostics and treatment of mycotic lesions in ear, nose, and throat. Vestn Otorinolaringol. 2012; 6:76-81.
[3] Liu J, Sun H, Wang S. Study on candida infections in intensive care unit from 2008 to 2012. Zhonghua Liu Xing Bing Xue Za Zhi. 2014; 35(3):326-8.
[4] Morales DK, Hogan DA. Candida albicans Interactions with Bacteria in the Context of Human Health and Disease. PLoS Pathog. 2010; 6(4):e1000886.
[5] De Pascale G, Antonelli M. Candida colonization of respiratory tract: to treat or not to treat, will we ever get an answer? Intensive Care Med. 2014; Jul 1.
[6] Roux D, Gaudry S, Khoy-Ear L. et al. Airway fungal colonization compromises the immune system allowing bacterial pneumonia to prevail. Crit Care Med. 2013; 41:e191-e199.
[7] Williamson DR, Albert M, Perreault MM. et al. The relationship between Candida species cultured from the respiratory tract and systemic inflammation in critically ill patients with ventilator-associated pneumonia. Can J Anaesth. 2011; 58:275-284.

Studies on biocidal properties of textile materials modified by organosilicone compounds

J. Walentowska[*,1], **J. Foksowicz - Flaczyk**[1], **M. Przybylak**[2] **and H. Maciejewski**[2]

[1]Institute of Natural Fibres & Medicinal Plants, Wojska Polskiego 71b, 60-630 Poznan, Poland
[2]Poznan Science and Technology Park - Adam Mickiewicz University Foundation, Rubiez 46, 61-612 Poznan, Poland
*Corresponding author: e-mail: judyta.walentowska@iwnirz.pl

The paper presents the research conducted to increase the resistance of cotton fabric modified by organosilicone compounds to action of moulds. Tetraethoxysilane (TEOS) with quaternary ammonium salts (QASs) modifiers, silver ions and triclosan, and octylotriethoxysilane with silver ions were used for modification of cotton fabric. The modified fabrics were tested for action of a mixture of 5 mould species, which most often cause decomposition of cellulose (*Chaetomium sp., Aureobasidium sp., Paecilomyces sp., Aspergillus sp., Penicillium sp.*). Evaluation of antifungal effect was done by determining the degree of moulds growth on the surface of tested fabric samples, the growth inhibition zone around the samples and the change of breaking force. Microscopic evaluation of the tested fabrics was also made with the use of SEM.

Keywords biocidal properties; silicon compounds; moulds; textile materials

1. Introduction

Textile materials containing natural fibres e.g. flax and hemp non-wovens, linen and cotton fabrics, find application both as finishing and insulation materials in automotive and construction industries and for air conditioning filters and medical materials. Problems occur when they are exposed to harmful external factors. High humidity and temperature, insufficient air circulation result in an enhanced growth of micro-organisms, especially mould fungi [1]. Uncontrolled mould fungi growth leads to complete degradation of cellulose, the main component of natural textile materials. For textiles used as medical materials is also important to obtain antibacterial properties. Among the many studies conducted by scientists from different countries very good antibacterial activity were obtained in the case of cotton fibers coated by silver nanoparticles with hexadecyltrimethoxysilane. Also this modification led to superhydrophobic cotton textiles [2]. Endowing textile materials with biocidal properties is an element of their multifunctional character. The antimicrobial agents used for finishing of textile materials include for example quaternary ammonium compounds, silver and gold nanoparticles, triclosan, N-halamines, silanes, polysiloxanes bearing quaternary ammonium salt groups, polybiguanides [3], [4], [5], [6], [7].

2. Materials and methods

2.1 Test fabric

100 % cotton fabric characterized with surface mass at 158 g/m^2. Before the modification process the fabric was bleached in a hydrogen peroxide (H$_2$O$_2$) bath.

2.2 Antimicrobial agents

Tetraethoxysilane (TEOS) with QASs modifiers labelled as M2 (iodide N, N-didecyl-N-methyl-N-(trimethoxysilylpropyl) ammonium and labelled as M8 were applied by padding method at 5 % concentration respectively of isopropanol and methanol solution for 15 minutes. The reference samples were treated also with isopropanol or methanol solution. TEOS with triclosan was applied by padding method at 10 % concentration in relation to TEOS of isopropanol solution (in acidic medium) for 15 minutes. The reference samples were treated also with isopropanol solution (in acidic medium). TEOS and octylotriethoxysilane with silver ions were applied by padding method at a molar ratio of 1: 0.24 of 2-methoxyethanol solution for 15 minutes. The reference samples were treated also with 2-methoxyethanol solution.

2.3 Antifungal test

Determination of the resistance of tested fabrics to the moulds action was conducted according to EN 14119:2003 Standard. Agar medium was used with pH 6.0-6.5 of the following composition: solution of mineral salts (NaNO$_3$ 2.0 g; KH$_2$PO$_4$ 0.7 g; K$_2$HPO$_4$ 0.3 g; KCl 0.5 g; MgSO$_4$·7H$_2$O 0.5 g; FeSO$_4$·7H$_2$O 0.01 g; distilled water up to 1000 ml); agar 20.0 g and glucose 20.0 g. The modified and unmodified samples of cotton fabric were exposed to the action of mixture of the following moulds: *Aspergillus niger* van Tieghem, *Chaetomium globosum* Kunze, *Aureobasidium pullulans* (de Bary) Arnaud, *Paecilomyces variotii* Bainier and *Penicillium ochrochloron* Biourge. For each fungus the spore concentration was determined with the use of a Thom chamber at about 10^6 spores ml^{-1}. During preparation of the final spore mixture for inoculation, the same amounts of each spore suspension were mixed together. The tested samples were placed on agar medium and inoculated with a suspension of testing fungi. Incubation of tested samples in temperature 29 ± 1^{0}C and relative air humidity at 90 % was conducted for 4 weeks. After the tests, evaluation of biocidal properties was performed on the basis of visual assessment by determination of moulds growth degree and change of the breaking force.

The rating system for moulds growth was as follows: 0 - no visible growth evaluated microscopically, 1 - no visible growth evaluated with naked eye but clearly visible microscopically, 2 - growth visible with naked eye, covering up to 25% of tested surface, 3 - growth visible with naked eye, covering up to 50% of tested surface, 4 - considerable growth, covering more than 50% of tested surface, 5 - very intense growth, covering all tested surface.

2.4 Evaluation of strength properties

Determination of breaking force of modified fabrics using the strip method was done according to the EN ISO 13934-1:1999 Standard. The samples were conditioned at 65 ± 2% relative air humidity and 29 ± 1^0 C temperature for 24 h before determination of breaking force.

The relative loss of breaking force (expressed in %) caused by moulds action was calculated from the mean value of six samples, using the formula:

$$S = 100 - \frac{\overline{A}}{\overline{B}} x100 \tag{1}$$

where $\overline{A}$ is the arithmetical mean value of breaking force for all samples exposed to moulds action, $\overline{B}$ is the arithmetical mean value of breaking force for all samples not exposed to moulds action.

2.5 Scanning electron microscopy (SEM)

The microscopic evaluation of surface changes of modified and unmodified samples of cotton fabric before and after 4 weeks exposure to the moulds was carried out with the use of a HITACHI S-3400N scanning electron microscope (SEM), where samples were coated with a thin layer of gold before observation.

3. Results

The biocidal properties of cotton fabric modified by different silicon compounds were studied by visual evaluation of moulds growth in scale from 0 to 5 with the use of stereoscope microscope. Moreover, the loss of breaking force was evaluated.

The cotton fabric modified by octylotriethoxysilane with silver ions and tetraethoxysilane (TEOS) with triclosan showed the best antifungal activity - no visible moulds growth evaluated microscopically, growth inhibition zone around the samples and no loss of breaking force. The cotton fabric modified by tetraethoxysilane (TEOS) with silver ions showed insignificantly weaker results - no visible growth evaluated with naked eye, but clearly visible microscopically. Also loss of breaking force was no observed. The cotton fabric modified by tetraethoxysilane (TEOS) with quaternary ammonium salts (QASs) modifiers showed diversified results. The fabric modified by (TEOS) with QAS modifier labelled as M2 did not show antifungal activity - very intense growth, covering all tested surface and significant loss of breaking force. QAS modifier labelled as M8 used for modification the tested fabric show better antifungal activity - growth visible with naked eye, covering up to 25% of tested surface, however loss of breaking force was no observed. For comparison, the reference samples and unmodified control sample showed decomposition by moulds - very intense growth, covering all tested surface and significant loss of breaking force. Table 1 shows all results.

Table 1 The resistance of cotton fabric modified by silicon compounds to moulds action.

Biocides	Degree of moulds growth (scale from 0 to 5)	Inhibition zone (mm)	Loss of breaking force (%)
Tetraethoxysilane (TEOS) + QAS (M2)	5	-	6
Tetraethoxysilane (TEOS) + QAS (M8)	2	-	0
Tetraethoxysilane (TEOS) + Triclosan	0	2	-4
Tetraethoxysilane (TEOS) + Silver ions	1	2	-1
Triethoxysilane + Silver ions	0	7	-4
Reference samples	5	-	8
Control samples	5	-	15

The results of antifungal tests are visible in Fig. 1. Visual evaluation shows that cotton fabric modified by tetraethoxysilane (TEOS) and octylotriethoxysilane with silver ions is not destroyed by moulds action - Fig. 1a) and Fig. 1b). Figure 1c) shows very intense growth of moulds, covering all tested surface of unmodified cotton fabric sample. This is also confirmed by SEM images which show structure of cotton fibres and occurrence of mould spores. Figure 2a) and Figure 2b) demonstrate that on the sample of cotton fabric modified by tetraethoxysilane (TEOS) with silver ions and octylotriethoxysilane with silver ions there is no moulds growth. Also no changes in the structure and on the surface of cotton fibres elementary were observed. Figure 2c) demonstrates very intense growth of mould fungi, covering all tested surface of unmodified cotton fabric sample. The structure and surface of cotton fibres elementary were destroyed in result of the exposure to moulds.

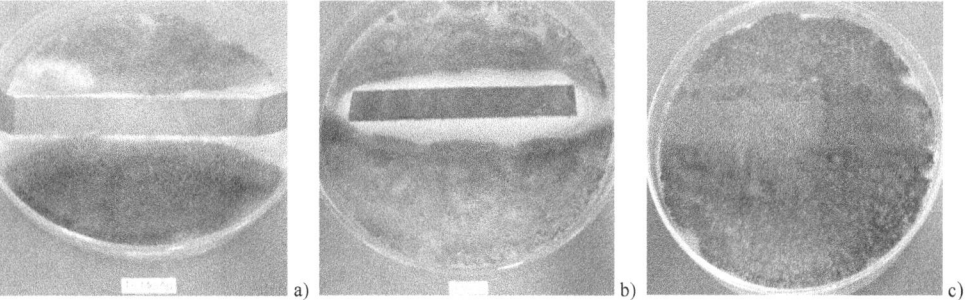

Fig. 1 Moulds growth on the agar medium covered with cotton fabric samples: a) modified by tetraethoxysilane (TEOS) with silver ions, b) modified by octylotriethoxysilane with silver ions, c) unmodified

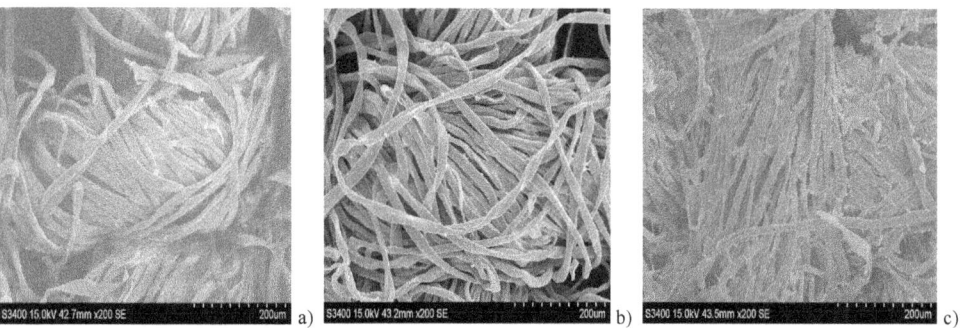

Fig. 2 SEM images of cotton fabric after moulds action: a) modified by tetraethoxysilane (TEOS) with silver ions, b) modified by octylotriethoxysilane with silver ions, c) unmodified

4. Conclusions

Among the antimicrobial agents applied to modification of cotton fabric the best results were obtained for cotton fabric modified by octylotriethoxysilane with silver ions, tetraethoxysilane (TEOS) with triclosan and tetraethoxysilane (TEOS) with silver ions, where the growth on agar medium around the sample was inhibited. The values of tensile strength for cotton fabric modified by silicon compounds for samples before and after tests were without any changes.

The achieved results will enable to continue the studies further in order to determine the washing durability of cotton fabric modified by selected silicon compounds and to confirm their biocidal properties.

Acknowledgments Study has been carried out within the SILANTEX Project - "Novel organosilicone compounds for gentrify natural fibres and textiles", financed by the National Research and Development Centre for Applied Research Program, 2012-2015.

References

[1] Angelini P. Filter Media: Using antimicrobial treated media in the pool or spa. Filtration+Separation 2007, January/February, 31-34.

[2] Xue Ch-H, Chen J, Yin W, Jia S-T, Ma J-Z. Superhydrophobic conductive textiles with antibacterial property by coating fibres with silver nanoparticles. Applied Surface Science. 2012; 258: 2468-2472.

[3] Scholz J, Nocke G, Hollstein F, Weissbach A. Investigations on fabrics coated with precious metals using the magnetron sputter technique with regard to their anti-microbial properties. Surface & Coating Technology. 2005; 192: 252-56.

[4] Tomšič B, Simončič B, Orel B, Žerjav M, Schroers H, Simončič A, Samardžija Z. Antimicrobial activity of AgCl embedded in a silica matrix on cotton fabric. Carbohydrate Polymers. 2009; 75:618-26.

[5] Windler L, Height M, Nowack B. Comparative evaluation of antimicrobials for textile applications. Environment International. 2013; 53: 62-73.

[6] Muñoz-Bonilla A, Fernandez-Garcia M. Polymeric materials with antimicrobial activity. Progress in Polymer Science. 2012; 37: 281-339.

[7] Fouda M. M.G, Abdel-Halim E.S, Salem S. Al-Deyab. Antibacterial modification of cotton using nanotechnology. Carbohydrate Polymers. 2013; 92: 943-954.

The planktonic Escherichia coli K-12 metabolic regulatory logic of glucose CRP-dependent repression is maladaptive for macrocolony biofilms formation: implications for the development of anti-biofilms drugs

José María Gómez Gómez

OAS-BioAstronomy Group. Camino Valparaiso S/N.45621 Segurilla (Toledo) Spain
E-mail: chemaseg@yahoo.es

Bacterial biofilms are multicellular sessile communities encased in a self-secreted extracellular matrix showing increased resistance to antibiotics compared with its planktonic counterparts. Hence, the development of new anti-biofilms drugs capable to inhibit the formation of bacterial biofilms is an active research front line. A key point in this investigation is the identification of potential new bacterial targets affecting to vital processes required for biofilm formation. A good experimental model where is possible to explore these issues are *Escherichia coli* K-12 macrocolony biofilms. Here, I present experimental evidence indicating that D-(+)-glucose suppresses via the c-AMP receptor protein (CRP) the formation of *E. coli* K-12 macrocolony biofilms on 1.5% (15 g/L) ABE hard agar surfaces. These results indicate that the regulatory logic of glucose CRP-dependent catabolic repression is maladaptive in these particular environmental and nutritive conditions; suggesting additionally that regulatory activity of cAMP-CRP complex to promote *E. coli* biofilm formation could be a new anti-biofilms drugs target.

Keywords: *E. coli* macrocolony biofilms; glucose repression; CRP-mediated catabolite repression; maladaptation; anti-biofilms drugs; ABE, agar bacteriológico europeo.

Introduction. Biofilms are complex communities of sessile, microbial cells encased in a self-secreted extracellular matrix that is composed primarily of extracellular polymeric substances (EPS): exopolysaccharides, proteins, and nucleic acids attached as whole on a wide range of biotic and abiotic surfaces [1], in natural as well as industrial and hospital settings [2]. For example, in *Escherichia coli* bacteria the extracellular matrix is formed by a variety of extracellular polymeric substances (EPS) and adhesions: type-1 pili, amyloid fibers (curli) and exopolysaccharides: cellulose, colanic acid, β-1,6-*N*-acetyl-D-glucosamine polymer (PGA) (reviewed in reference [3]). Many clinically relevant antibiotics are ineffective in the treatment of biofilm-related bacterial infections because bacteria in biofilms can withstand antibiotic treatment [4,5]. Such antibiotic tolerance is a matter of medical concern especially because the emergence of an increasing number of multi-drug resistant strains is not accompanied of discovery of new antimicrobial agents [6]. Hence, it is fundamental to understand the molecular basis of formation, maintenance, and dispersal of bacterial biofilms in order to identify novel targets that could potentially be used for inhibiting the formation of these bacterial multicellular aggregates [6].

The decision to transition between a planktonic and a biofilm sessile life-style is orchestrated in the *E. coli* bacterial cells through the modulation the production of near ubiquitous bacterial second messenger cyclic-di-GMP (c-di-GMP), a dinucleotide known to modulate many different aspects of bacterial physiology [7]. In addition, other cyclic nucleotide important to bacterial biofilm formation is the cyclic-AMP (cAMP) [8]. Catabolite repression (CR) is the preferential utilization of glucose as a carbon source by bacteria [9,10]. When glucose is available in the medium, uptake and utilization of alternative carbon sources are repressed [9, 10]. In *E. coli*, the availability of glucose in a nutritive medium prevents the cyclic AMP (cAMP) synthesis by inhibition of the cAMP-producing enzyme cyclase (Cya) activity; hence the intracellular levels of cAMP, and consequently the DNA binding activity (which is controlled allosterically by cAMP) of its intracellular effector, the cAMP receptor protein CRP, are high when poor carbon sources are available and low when glucose is present as nutrient. Thus, through this regulatory molecular mechanism, the glucose suppresses the CRP-mediated activation of genes necessary for growth in secondary carbon sources [9,10]. This traditional view of the catabolite repression in *E. coli* have been recently completed, it has been suggested that the information that is transferred to CRP is not the general carbon availability, but rather the balance between carbon catabolism and the capacity for anabolism [11]. Thus, it has been reported that the inhibition of Cya by α-ketoglutarate and related α-ketoacids provide a negative-feedback loop that link both metabolic processes closing the regulatory circuit between carbon availability and CRP activity [11]. However, despite extensive studies on this cAMP signalling pathway, a number of important issues remain unanswered even for this well-known system, in part because the study of catabolic repression have been done mainly in planktonic cultures.

Recently, it has been reported that wild-type *E. coli* K-12 strains develop macrocolony biofilms with volcano-like morphology after 14 days of growth at 37°C on semisolid 0.6 % (6g/L) Agar Bacteriológico Europeo

(ABE) surfaces [12]. Additionally, it was determined that glucose represses the formation of this macrocolony biofilms via the cyclic AMP (cAMP) receptor CRP protein, identifying this as an important regulatory factor of the development of this kind of biofilms [12]. During the study of the effect of glucose upon development of E. coli K-12 colonies in higher ABE agar concentration 1,5% (15g/L) it was observed unexpectedly that the D-(+)-glucose 0.5% (5g/L) suppressed the formation of macrocolony biofilms on this agar surface (Fig. 1A). To gain information if the effect of glucose in the development of this kind of biofilms was mediated via the activity of the cAMP-CRP complex, the effect of a Δcrp mutation on macrocolony biofilm formation was assayed. The figure 1B shows the appearance of the macrolony (morphotype) that developed a wild type E. coli MG1655 strain on a solid 1.5% (15 g/L) ABE agar surface and their isogenic strain GSO549 harbouring the Δcrp::cat chromosomal mutation [12] after 14-day-old of growth at 37°C. While the MG1655 wild-type strain developed in this kind of the hard ABE agar surfaces a 3-D complex macrocolony biofilm (Fig. 1B and 1C) (to my knowledge this kind of E. coli morphotype in such high agar concentration has not been reported before) the strain lacking of CRP transcriptional regulator (GSO549) showed a dramatic impairment in its ability to form this kind of macrocolony biofilm (Fig. 1B), having developed its colony a size similar to observed when this Δcrp mutant strain was grown with 0.5 % D-(+)-glucose as additional nutrient of LB medium (to compare Fig. 1A versus 1B).

Taken together these results indicate that the formation of the E. coli K-12 macrocolony biofilms on surfaces prepared with 1.5% ABE agar is repressed by glucose via the regulatory activity of the cAMP-CRP complex. Intriguingly, the macrocolony of the wild-type strain shown in Fig. 1B and 1C has a big unexpected size; it is tempting to speculate with the possibility that this macro-colonial size was achieved no only through of canonical growth by binary fission, but also by a kind of CRP controlled colonial "spreading" movement. This possibility will be explored in future experiments. Interestingly, it has been estimated to this respect for instance that cAMP-CRP complex regulates directly from a minimum of 378 to maximum of 500 genes in E. coli, with other functions supposedly not related directly with the metabolism [13], for example the control of production of different cryptic fimbriae [14].

In principle, theoretically given the metabolic regulatory logic exhibited by the cAMP and CRP regulatory circuit in liquid medium, the glucose inhibition of the biofilm formation should not be expected because the glucose is the carbon source preferred by E. coli. Thus, the effect of the glucose on macrocolony biofilms formation under the environmental and nutritional conditions shown in this work (Fig. 1A and Fig. 1B) is contra-intuitive and apparently not adjustable to the anabolic/catabolic logic of CRP-dependent circuitry that work in E. coli when exhibit a planktonic lifestyle [10,11]. Under this logic it should be expected that glucose should stimulate the growth and development of E. coli colonies and that it does not to cause the inhibition of these bacterial processes. In other words, the glucose effect in this case is maladaptive, i.e., by using glucose in this condition of growth the E. coli bacteria cells paradoxically loss biological fitness and adaptability.

What is the underlying logic of the glucose regulation of biofilm formation, i.e., what are the physiological reasons that explain this glucose inhibitory effect? In others words, if the priority for E. coli is to use glucose as carbon source, why does glucose inhibit the macrocolony biofilm formation? Indeed, this is an apparent intriguing paradox that requires of more experimental work to understand it.

It is well known that when D-(+)-glucose 0.5 % (5g/L) is added to Luria Bertani (LB) medium harnessed with Difco agar 0.5 % (5g/L) E. coli swarming behaviour on semisolid agar surfaces [15]. Although how glucose promotes swarming motility is still poorly understood [16]. That is, the effect of glucose on the E. coli behaviour apparently depends on the agar concentration: inhibitory to a high concentration, but stimulatory in low concentration. It has been previously reported a glucose repressive effect mediated in part by cAMP and CRP [17] on biofilm formation in Escherichia coli under different conditions of growth, but the physiological logicality of this suppressive glucose effect was not properly discussed [17]. Although it is clear that more experiments are necessary to understand why the planktonic metabolic logic of the CRP-dependent catabolic repression is maladaptive for the biofilms development.

Regardless of how finally this paradox will be explained, the inhibitory effect of glucose and the regulatory activity of cAMP-CRP on E. coli biofilm formation could be used as new anti-biofilms drugs targets.

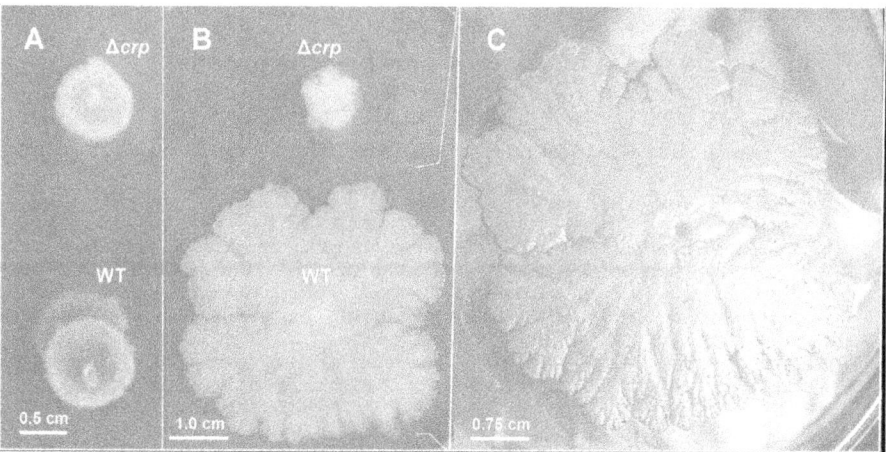

Fig. 1 *The formation of E. coli K-12 macrocolony biofilms 14-day-old in ABE agar surfaces 1.5% (15g/L) is under glucose CRP-dependent repression.* Two overnight cultures of wild-type (WT) MG1655 *crp+ E. coli* K-12 strain and their isogenic MG1655 *Δcrp::cat strain* (GSO549) grown in Luria Bertani (LB) medium [12] were used to inoculate with a toothpick semisolid surfaces prepared with Agar Bacteriológico Europeo (ABE) 1.5% (15g/L) containing LB medium as nutrient (A and B) in Petri dishes. D-(+)-glucose 0.5% (5g/L) was added to LB medium in (A). The Petri dishes were sealed with parafilm® to avoid the water evaporation and were incubated 14 days at 37°C, after this incubation time, the developed colonies were photographed with reflected light. **(A)** The D-(+)-glucose has a strong inhibitory effect on macrocolony biofilms development. **(B)** The cAMP receptor protein (CRP) is required for the formation of *E. coli* K-12 macrocolony biofilms on these hard ABE agar surfaces. **(C)** Enlarged view of *E. coli* K-12 wild-type macrocolony biofilm shown in (B).

References

[1] López D, Vlamakis H, Kolter R. Biofilms. Cold Spring Harbor Perspectives Biology. 2010; 2:a000398. doi: 10.1101/cshperspect.a000398.

[2] Hall-Stoodley L, Costerton JW, Stoodley P. Bacterial biofilms: from the natural environment to infectious diseases. Nature Reviews Microbiology. 2004; 2:95-108.

[3] Beloin C, Roux A, Ghigo JM. *Escherichia coli* biofilms. Current Topic in Microbiology and Immunology. 2008; 322:249-89.

[4] Mah TF, O'Toole GA. Mechanisms of biofilm resistance to antimicrobial agents. Trends in Microbiology. 2001; 9:34-9.

[5] Kostakioti M, Hadjifrangiskou M, Hultgren SJ. Bacterial biofilms: development, dispersal, and therapeutic strategies in the dawn of the postantibiotic era. Cold Spring Harbor Perspectives Medicine. 2013; 3:4a010306. doi: 10.1101/cshperspect.a010306.

[6] Spellberg B, Powers JH, Brass EP, Miller LG, Edwards Jr JE. Trends in antimicrobial drug development: implications for the future. Clinical Infectious Diseases. 2004; 38:1279-86.

[7] Hengge R. Principles of c-di-GMP signalling in bacteria. Nature Reviews Microbiology. 2009; 7:263-73.

[8] Karatan E, Watnick P. Signals, regulatory networks, and materials that build and break bacterial biofilms. Microbiology and Molecular Biology Reviews. 2009; 73:310-47.

[9] Botsford JL, Harman JG. Cyclic AMP in prokaryotes. Microbiological Reviews. 1992; 56:100-22.

[10] Görke B, Stülke J. Carbon catabolite repression in bacteria: many ways to make the most out of nutrients. Nature Reviews Microbiology. 2008; 6:613-24.

[11] You C, Okano H, Hui S, Zhang Z, Kim M, Gunderson CW, Wang YP, Lenz P, Yan D, Hwa T. Coordination of bacterial proteome with metabolism by cyclic AMP signalling. Nature. 2013; 500:301-6.

[12] Gómez Gómez J, Amils R. A novel cellular autoaggregative developmentally CRP regulated behaviour generates massively chondrule-like formations over surface of old *Escherichia coli* K-12 macrocolony biofilms. Advances in Bioscience and Biotechnology. 2014; 5:727-39.

[13] Shimada T, Fujita N, Yamamoto K, Ishihama A. Novel roles of cAMP receptor protein (CRP) in regulation of transport and metabolism of carbon sources PLoS ONE. 2011; 6:e20008. doi: 10.1371/journal.pone.0020081.

[14] Korea, CG, Badouraly R, Prevost MC, Ghigo JM, Beloin C. *Escherichia coli* K-12 possesses multiple cryptic but functional chaperone-usher fimbriae with distinct surface specificities. Environmental Microbiology. 2010; 12:1957-77.

[15] Gómez-Gómez JM, Manfredi C, Alonso JC, Blázquez J. A novel role for RecA under non-stress: promotion of swarming motility in *Escherichia coli* K-12. BMC Biology. 2007; 5:14. doi:10.1186/1741-7007-5-14.

[16] Girgis HS, Liu Y, Ryu WS, Tavazoie S. A comprehensive genetic characterization of bacterial motility. PLoS Genetics. 2007; 3:1644-60.

[17] Jackson DW, Simecka JW, Romeo T. Catabolite repression of *Escherichia coli* biofilm formation. Journal of Bacteriolgy. 2002; 184:3406-10.

The toxicity of the fluorides in oral hygiene products

K. Peros[*,1]

[1] Department of Pharmacology, School of Dental Medicine University of Zagreb, Salata 11, 10000 Zagreb, Croatia
*Corresponding author: e-mail: peros@sfzg.hr, Phone: +385 1 4590211

Through recent decades, fluoride caused the uncertainty between dentists and toxicologists due to its favourable caries preventive effects and deleterious effects at high concentrations in humans suffering from fluorosis. The purpose of this article is to present review and practical findings that focus on caries preventive fluoride preparations with respect to potential toxicological implications. This article presents an overview of the current research on the toxicology of fluoride exposure with emphasis on fluoride preparations for oral hygiene intended for home use - self administer use. The goal of this article is to enhance necessarily of understanding of the dose of fluoride in preparations for oral hygiene which may affect physiological processes in humans.

Keywords fluoride; toothpaste; toxicity; prevention

1. Introduction

Fluorine (F) is a halogen atom which has a compact structure, distinct electronegative properties and high reactivity. In the gaseous state it is almost colorless, in thick layers shows green and yellow colour, has sharp and stimulus odor, reminiscent of chlorine and ozone. It is made by electrolysis of anhydrous liquid hydrogen fluoride. Due to the high reactivity, it is not possible to find fluorine in natural state. It forms stable compounds with all elements except nitrogen and noble gases. Bonded with hydrogen shows high affinity for silica and silicates. Fluorides are salts of hydrofluoric acid. All fluorides are soluble in water, except calcium fluoride and magnesium fluoride.

People are exposed to fluorides in the air, food and water they consume as well as dental preparations used in the prevention of dental caries. Absorption of fluorides depends on the route of their administration: by the respiratory tract - inhalation of gas or dust with fluorides released into the atmosphere from various industrial plants (factories related to superphosphate, aluminum, steel, glass, ceramics); through the skin - contact (eg. workers in an industry that uses fluoride) or by the digestive system (food intake, water). The largest part of fluoride introduced into the human body is through digestive system, where it is absorbed by simple diffusion process in the stomach and small intestine, as undissociated hydrogen fluoride. The rate of absorption, as with other weak acids, depends on the pH of the environment in the stomach and small intestine. Fluoride uptake in the gastrointestinal tract can be enhanced by iron ions, phosphates and sulfates, and reduced by ions of aluminum, calcium and magnesium which form stable complexes with fluorine ions. Fluoride concentration in plasma varies depending on the amount of the fluoride intake, without homeostatic regulation. Fluoride is distributed to all tissues and organs and accumulates in calcified tissues. More than 50% of fluoride entered the body is retained in the skeleton and teeth, containing 99% of the total fluoride in the body. Fluorides are excreted from the body by the kidneys - by glomerular filtration, rapidly - 20% of the initial amount can be found in the urine after 3 hours. The amount of fluoride in urine is a reflection of the daily intake, total intake of fluoride, the form in which the fluoride is ingested, and the health status of human, especially kidney function. The half-life is between 2 and 9 hours. Renal clearance of fluoride in adults is 30 to 50 ml / min. The rate of excretion of fluoride is very important in poisoning.

2. Fluoride in preparations for oral hygiene

At the end of the nineteenth century the correlation of exposure to fluoride and reduced incidence of dental caries was observed. The effects of fluoridation of drinking water on the incidence of dental caries in children shown in "Newburgh-Kingston" study, led to the systematic addition of fluoride to drinking water in public water supplies in some countries. Croatia was not among those countries. Other measures of endogenous fluoridation included the addition of fluoride in salt or milk, or the fluoride given in the form of tablets. In Croatia, as well as many other countries, the endogenous fluoridation was only available in tablet form. Many studies have shown effects of fluoride applied directly to the tooth surface and at the mid-twentieth century adding fluoride in preparations for oral hygiene began. Study conducted on island Lewis' showed correlation of reduction in dental caries to usage of fluoride containing toothpaste. According to the World Health Organization, fluoride toothpastes are used by over 500 million people in the world today, and among them we can find the Croatian residents. The most common anticaries active compounds of toothpaste are fluorides - just as fluoride, sodium fluoride, tin fluoride (stannous fluoride), sodium phosphate monofluorides and aminofluorides. Such pastes, regardless of which fluoride is added, are providing a similar reduction in caries. The most widespread is the use of monofluoride phosphate because of good compatibility with other ingredients of toothpaste. In Croatia, toothpastes containing sodium fluoride are prevailing. Toothpastes designed for use in adults, usually contain 1450ppm of fluoride or 0.32% of sodium fluoride. Toothpastes designed for use in children 6 to 12 years, usually contain 1000ppm of fluoride, or 0.2% of sodium fluoride. Toothpastes designed for use in children younger than 6 years, usually contain 500ppm of fluoride, or 0.1% of sodium fluoride. Fluoride is often added to the mouthwash, which usually contains sodium fluoride at a concentration of 0.02 to 0.05% and is intended for use in adults. Mouthwash is not intended for use in children under 6 years of age. Some mouthwashes often have ethyl alcohol as a component in a concentration between 14 and 27%. Although ethyl alcohol contributes in reduction of dental plaque, there are some doubts about adding of alcohol to oral hygiene products.

3. Doses

The lethal dose for man is from 32 to 64 milligrams of fluorine per kilogram of body weight. A lethal dose of sodium fluoride is 5 to 10 grams for men weighting 70kg. In children even a half of a gram of sodium fluoride can be fatal. In 100 grams of toothpaste for adults there are 0.32 grams of sodium fluoride. So, for a man who weights 70kg lethal dose of sodium fluoride is contained in about 1500 grams of toothpaste for adults. For a child who weights 20kg lethal dose of sodium fluoride is contained in about 440 grams of toothpaste for adults, ie from 700 to 1400 grams of toothpaste for children. Mouthwashes are usually packaged in vials of 200-500ml size and containing from 0.1 to 0.25 grams of sodium fluoride in the packaging. Thus, for a man who weights 70kg lethal dose of sodium fluoride is contained in 20 vials per 500ml of mouthwash. For a child who weights 20kg lethal dose of sodium fluoride is contained in about 5.5 bottles per 500ml of mouthwash.

4. Acute poisoning

Symptoms of acute poisoning that require immediate medical attention may be induced by dose of 5 mg fluoride / kg body weight. Among fluoride preparations for oral hygiene, such poisoning may be consequence of intentional or accidental ingestion of large quantities of fluoride toothpaste at a time (for a man who weighs 70 kg this dose is contained in 240 g of toothpaste containing sodium fluoride) or drinking large quantities of mouthwash at once (for a man who weights 70kg this dose is contained in 1540 ml of a mouthwash with sodium fluoride). Preparations intended for endogenous application for the prevention of dental caries, as tablets of sodium fluoride, can also cause poisoning. Such poisonings are common in young children who have fluoride tablets within reach. There are described cases of acute poisoning caused by application of fluoride solutions for professional topical fluoridation (eg. 4% solution of tin fluoride) as well.

After swallowing the mentioned (or greater) amount of fluoride preparations for oral hygiene, hydrofluoric acid occurs in stomach and affects as local irritant on the mucous membranes of the digestive tract causing nausea, vomiting, diarrhea, and abdominal pain. After absorption, there are noticeable effects on enzyme systems, and on the cardiovascular, respiratory and central nervous systems as well, in the form of hypocalcemia, cardiac arrhythmias, paresthesia, carpopedal spasm, tonic clonic seizures and cramps. The blood pressure decreases, due to central vasomotor depression, and direct action on the myocardium. Respiratory center is at first stimulated, then depressed. Patients show signs of increased irritability of the central nervous system with paresthesia. Death occurs due to central vasomotor depression and respiratory failure, usually after 3 to 4 hours.

4.1 Emergency treatment

In the treatment of acute poisoning, it is important to remove unabsorbed fluoride from the digestive system and to hasten the elimination of already absorbed part. Gastric lavage with 0.15% solution of calcium hydroxide or 1% solution of calcium chloride should be conducted, because calcium will bound fluorine and prevent its absorption. In the home setting, chalk can also be used as well as inducing vomiting and then drinking milk. As a laxative 30 g of a mixture of equal parts of sodium sulfate and magnesium oxide is used, as magnesium also binds fluorine. Alkaline diuresis may hasten the elimination of fluoride. To resolve tetany, 10 ml of 10% solution of calcium gluconate is given intravenously, if necessary every 15-30 minutes. Loss of NaCl should be recovered by the continuous intravenous infusion of equal parts of isotonic saline and glucose.

5. Chronic poisoning

Fluoride chronic poisoning occurs with intake of 10 to 15 mg of fluoride per day, during long period of time. This amount is contained in in the form of sodium fluoride in the 7 to 10 grams of toothpaste containing sodium fluoride for adults, or in about 50 ml of a mouthwash with sodium fluoride. Also, chronic poisoning may be due to improper intake of fluoride tablets, prolonged use of drinking water with a fluoride concentration from 2 mg F / L, or occupational exposure to amount of 10 to 15 mg of fluoride per day, by inhalation or skin contact as well as through mucous membranes.

Excessive use of fluorides is reflected in the stomach and kidneys. Changes in the stomach are characterized by structurally and functionally reversible damage to cells of the gastric mucosa. In kidneys, the ability to concentrate urine is decreased, which can lead to disturbance of electrolyte balance. The most prominent change in fluoride chronic poisoning is evident in mineralized tissues, as bones and teeth. Bones become osteosclerotic, with increased density and calcification. Arthralgia and limitations of mobility of the cervical and lumbar spine are present. The degree of change in bone varies from barely noticeable to numerous radiological egzosthosis along the skeleton, calcification of ligaments, tendons and muscle insertions to bone. Secondary, anemia may occur due to the reduction of bone marrow. Changes in the appearance of the teeth caused by excessive intake of fluoride during tooth mineralization (from the fifth month of intrauterine life to about eight years of age) are called dental fluorosis. A typical appearance is with white opaque areas in the form of stripes of different widths on enamel of sets of homologous teeth. In severe fluorosis complete teeth are milky white as chalk. Opaque areas are porous (hypomineralised) and after the eruption of teeth, depressions (pits, wells) and brown discolorations can occur. The wells are the result of posteruptive cracking of enamel. Fluorotically mineralized enamel surface is very easy to break by the mechanical shock, since it includes hypomineralised lesions. Dental fluorosis develops usually at permanent dentition, rarely in primary teeth, usually on the labial surface of the upper and lower incisors. Dental fluorosis may commonly occur in children due to high ingestion of fluorides during (non) professional topical fluoridation or inappropriate ingestion of toothpaste when brushing teeth. The daily dosage of between 0.03 and 0.10 mg of fluoride per kilogram of body weight can lead to dental fluorosis. For a child that weights 15kg this dose is contained in 0.31 gram of toothpaste for adults. For upper incisors critical age for the development of fluorosis is between one and two and a half years of age. After six years of age the risk of dental fluorosis is insignificant.

5.1 Treatment

It is necessary to reduce the intake of fluoride in the body. Digestive disorders and kidney function as well as joint pain and limitations in mobility, are prone to spontaneous recovery after stopping intake of fluoride. Dental changes acquired in childhood are permanent. The only aesthetic therapy is fixed-prosthetic restorations.

6. Conclusion. Poisoning prevention

The victims of fluoride poisoning from oral hygiene products are, mostly, children. It is necessary to keep the oral hygiene products out of the reach of children. When your child is brushing teeth, it is necessary to supervise and to put a proper amount of toothpaste on the toothbrush, and the tube with the rest of the paste immediately return to the place beyond the child's reach. It is not recommended that a household have stock (more tubes or bottles) of oral hygiene products. The same measures are also needed for people with developmental disabilities and in some psychiatric patients. For the prevention of chronic poisoning, it is important that among the available modes of fluoridation (endogenous, professional topical, fluoridated toothpaste) select and implement only one.

Acknowledgements The support of Croatian Ministry of science, education and sports is gratefully acknowledged (MZOS grant No. 065-0650445-0406, grant holder: prof. dr. sc. Kata Rosin-Grget).

References

[1] Rosin-Grget K. Fluorides. In: Lincir I. et al. Pharmacology for dentists. Zagreb: Medicinska naklada; 2011. p. 333-54.
[2] Browne D, Whelton H, O'Mullane D. Fluoride metabolism and fluorosis. Journal of Dentistry. 2005;33:177-86.
[3] Petersen PE, Lennon MA. Effective use of fluorides for the prevention of dental caries in the 21st century: the WHO approch. Community Dentistry and Oral Epidemiology. 2004;32:319-21.
[4] Whitford GM. Acute and chronic fluoride toxicity. Jornal of Dental Research. 1992;71:1249-54.
[5] Whitford GM. Intake and metabolism of fluoride. Advances in Dental Research. 1994;8:5-14.
[6] Yeung CA. Systematic review of the efficacy and safety of fluoridation. Evidence Based Dentistry. 2008;9:39-43.

[7] Ast DB, Finn SB, and McCaffrey I. The Newburgh-Kingston Caries Fluorine Study. I. Dental Findings after Three Years of Water Fluoridation. American Journal of Public Health. 1950;40(6):716–24.

[8] Lincir I. Preparations for oral hygiene. In: Lincir I. et al. Pharmacology for dentists. Zagreb: Medicinska naklada; 2011. p. 355-63.

[9] Gerrie AS, Kotecha SA, Chesney AE, Rachlis A, Cheung MC. Fluorosis detected by trephine biopsy. British Journal of Haematology. 2009;144(3):278.

Using peptidoglycan as a basis for measuring murein-destroying activity of serum

V. Y. Ziamko[*,1], V. K. Okulich[2]

[1] Medical faculty, Vitebsk State Medical University, Frunze Str.,Vitebsk, Belarus
[2] Department of Microbiology, Vitebsk State Medical University, Frunze Str.,Vitebsk, Belarus
*Corresponding author: e-mail: torinet@tut.by, Phone: +375-29-146-07-99

Determining activity of enzymes that destroy peptidoglycan – an important part of the bacterial cell wall – can substantially help in diagnosis of bacterial infections. We propose a method of separation of a cell wall from gram-positive bacteria, that significantly simplifies and allows reducing the cost of the isolation of peptidoglycan, that can be used as substrate for measuring activity of such peptidoglycan-destroying enzymes.

Keywords: gram-positive bacteria; peptidoglycan; enzyme activity; Congo red.

Abbreviations:
AMP - antimicrobial peptides;
PG - peptidoglycan;
PLCR- peptidoglycan labeled with 2% solution of Congo red;
pkat – picokatals;
E_{op} – optical density.

Introduction

Two classes of substances – endogenous antimicrobial peptides (AMP) and endogenous antimicrobial proteins, play a very important role in protection of organism against microbial agents [1,2]. Antimicrobial proteins have rather large size containing more than 100 amino acids and most often show lytic enzymatic activity, ability to bind nutrients for bacterial cells or contain sites directed against specific microbial macromolecules. AMP have the smaller size and, as a rule, destroy the structure or the function of a cellular membrane of microorganisms [2,4].

Nowadays, there are hundreds of AMP found in epithelial tissues, phagocytes and biological liquids. Some AMP are synthesized constantly, while synthesis of others is induced in response to an infection or an inflammation. The majority of AMP is presented by cationic granules, associated peptides with affinity to components of a microbial cellular wall, for example peptidoglycan.

When monitoring a course of infectious and inflammatory disease in clinical laboratory practice elevation of the AMP levels can be useful as marker of system activation of neutrophils [7]. The knowledge of a role of enzymes, destroying peptidoglycan during infectious process, is very important for understanding deep mechanisms of interaction between microorganism and immune system of a macroorganism. That is necessary for development of new methods of diagnostics and treatment of infectious diseases.

In our opinion, there is scientific and practical interest in developing new methods of determination of serum activity of the peptidoglycan-destroying enzymes and in assessment of their activity in course of various pathological processes.

Purpose: to isolate peptidoglycan from cell wall of gram-positive bacteria and to analyze enzymatic activity of blood serum to destroy the isolated peptidoglycan in patients with purulent otitis.

Methods

1. Isolation of peptidoglycan from gram-positive bacteria.

Isolation of PG from cell wall of gram-positive bacteria was carried out by method proposed by V. Lvov, B. Pinegina and R. Khaitov in our modification[3]. *Microccocus luteus ATCC 10240* was used as a culture. The isolated PG, labeled with 2% solution of Congo red was used as a substrate for determining activity of enzymes, that destroy PG [7,8]. The quality of the received PG has been determined by means of confocal microscopy [Figure 1].

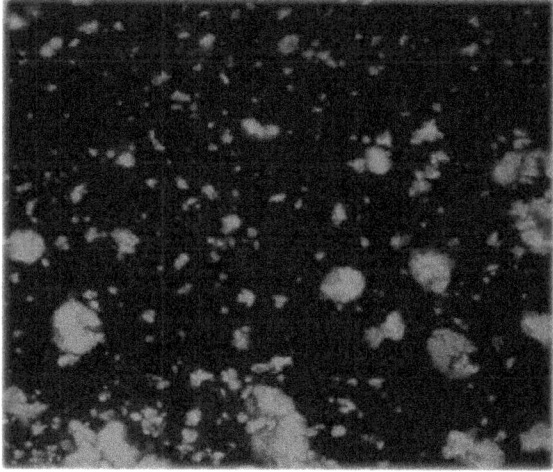

Figure 1 - Substrate - peptidoglycan labeled with Congo red. Diametre of particles from 2 to 9 microns (objective *40x*, zoom=5).

We didn't remove teichoic acids, as it is known that their removal changes the specificity of bacteriolytic enzymes.

Isolation of peptidoglycan is carried out as follows:

1. Bacteria were grown on meat-peptone agar. Overnight culture was washed twice with saline for removing components of the nutrient medium. It was centrifuged 1700 turns/min for 10 min, and 2000 turns/min for 40 min.

2. 20 ml of 30 % aqueous phenol pre-warmed to 90 °C was added to the resulting suspension. The mixture was being stirred for 20 minutes at a temperature of 65-68 ° C and cooled to 20 ± 2 °C and centrifuged 3 times for 10 min at 1700 turns/min, removing the supernatant each time. At this stage lipopolysaccharides (LPS), proteins, nucleic acids and other bacterial cell wall components, non-covalently associated with PG, were removed. 30 % aqueous phenol was used as an extractant at temperature of 65-68 °C (extraction by Westphal), since there is a complete homogenization of a mixture of phenol and water at this temperature.

3. Distilled water was added to a total volume of 300 ml and 3 ml of 100 % acetic acid. The product, obtained in the preceding step, was being stirred at 100 °C for 3 hours to get free from trace amounts of LPS.

4. After cooling the precipitate was placed on a magnetic stirrer for 10 minutes and separated by centrifugation for 20 minutes at 2000 turns/min, washing it 3 times with water after each centrifugation.

5. Then dialysis was carried out at 21±2C°, with removing the dialysate 3 times (1 per day) in a vessel and adding fresh buffer solution. Dialysis was performed to remove low molecular weight products that were formed during hydrolysis. 0.05 M sodium acetate solution pH 5.8 was used as a buffer.

6. The suspension, labeled with 0.5 % solution of Congo red, was washed twice with saline and centrifuged at 1000 turns/min for 75 minutes after each wash to remove unbound solution of Congo red [6].

Qualitative and quantitative characteristics of the isolated peptidoglycan were evaluated by confocal microscopy. Propidium iodide was added to the reaction mixture in the concentration of 20 g/ml. Further, preparation "hanging drop" was prepared. Stratified scanning was carried out immediately on confocal microscope Leica TCS SPE.

Scanning was performed in XYZ regimen with the use of 2 consequential spectral channels, whose characteristics were selected so there was no mutual overlap of the fluorescence from different dyes. The first channel (488 nm laser, spectral detection area of 500-530 nm) was used for excitation and detection of fluorescence FITC. The second channel (532 nm laser, spectral detection area is 570-650 nm) was used for propidium iodide. Scanning was performed with a resolution on the plane 150 nm, the distance between layers - 500 nm. The time interval between scans was 3 minutes (Figure 1).

2. Determining peptidoglycan - destroying activity of blood serum in patients with purulent-inflammatory diseases

18 blood sera were taken from patients with purulent-inflammatory diseases and 18 sera were taken from healthy donors. Patients with purulent-inflammatory diseases were divided in 2 groups: 10 people with acute purulent otitis and 8 – with chronic purulent otitis.

The serum was centrifuged 1.5 thousand/min. for 10 minutes. 300 µl of PLCR solution and 100 µl of serum were added into the first series of eppendorfs. 300 µl of PLCR solution and 100 µl of serum, pre-heated at 56 ° C during an hour in order to inactivate complement, were added into the second series of eppendorfs. Samples, containing Tris-HCl buffer at pH 7.4 in an amount of 300 µl and 100 µl of serum, served as control. All the samples were incubated in thermostat at t = 37 ° C for 24 hours. When enzymes in blood serum destroyed PG, Congo red became soluble, changed its color from colorless to red with a maximum spectrum of absorption at a wavelength of 495 nm. After incubation, the samples were removed from the thermostat and were centrifuged for 10 min (10 thousand/min; MICRO 120) for the remaining deposition of PLCR. 150 µl of the solution were taken from the supernatant in duplicate and transferred into the wells of 96-well polystyrene plate. The plate was placed in a multichannel spectrophotometer F300, where absorbance was determined in the wells in the wavelength of 492 nm.

Interjacent result was expressed in units of optical density and was calculated as the difference between the optical densities of the test samples and their corresponding controls.

To convert the result into pikokatals we used a formula, obtained after constructing the calibration graph for the breeding of Congo red, in which the enzyme activity dependence from the optical density of the solution was reflected, assuming that after splitting of one substrate molecule, 1 molecule of Congo red goes into solution (Figure 2).

$$Y = [-0,001 + 0,026 \times Eop] \times 9,921$$

Where Y - the desired result;

Eop - optical density of the sample minus optical density of control.

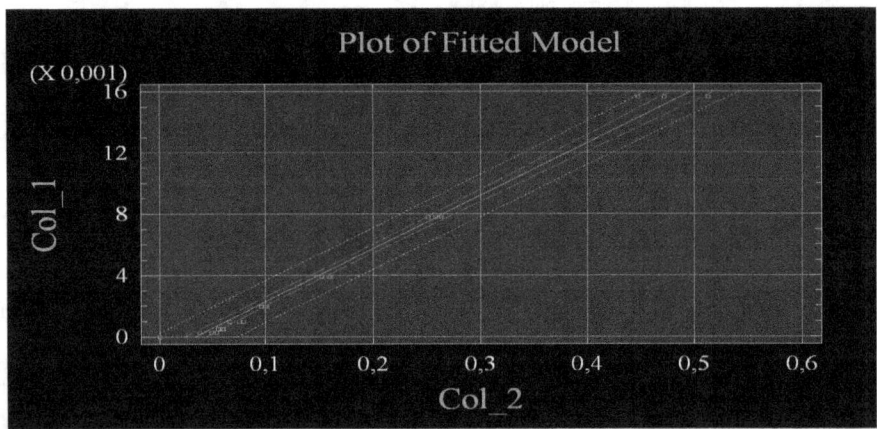

Notes: col_1 – concentration of Congo-red, mg/ml
col_2 – optical density of solution

Figure 2 - Graph of optical density of the solution concentration of Congo red

Since the analysis of the data distribution showed their nonparametric distribution, statistic processing was performed by using the Kolmogorov-Smirnov test. The differences were considered significant at $p < 0.05$.

Results and discussion

The experiment has revealed that peptidoglycan-destroying activity of blood serum in patients with purulent-inflammatory diseases was significantly higher in comparison with donors ($p < 0.05$). After complement inactivation ability of enzymes to destroy PG significantly reduced ($p < 0.05$). Results are presented in table 1.

Table 1. Peptidoglycan- destroying activity of blood serum before and after complement inactivation in donors and patients with purulent otitis

	Group	N	Median, pkat	Percentile, pkat	Significance of differences
1	Donors	18	0,087	0,079-0,088	P_{1-2}=0,03 P_{2-3}=0,02
2	Patients with purulent otitis	18	0,112	0,085-0,136	P_{1-3}>0,05
3	Patients with purulent otitis after complement inactivation	18	0,076	0,047-0,099	

In our opinion, the results show that assessment of activity of peptidoglycan-destroying enzymes can be useful in diagnostics and treatment of infectious diseases.

Conclusion

1. The described technique allows to obtain the peptidoglycan component from the cell wall of Gram-positive bacteria and to determine antimicrobial activity of blood serum according to its ability to destroy peptidoglycan, which is one of the mechanisms of nonspecific resistance, allowing macroorganisms to fight against infection.
2. It was found that enzyme ability to destroy peptidoglycan was significantly higher in patients with purulent-inflammatory diseases (0,112 pkat), than in donors (0,087 pkat, p <0.05)
3. After complement inactivation, enzyme ability to destroy peptidoglycan was significantly reduced (0,076 pkat, p<0,05)

Acknowledgements. We would like to thank F.I. Pleshkov, who contributed in the work and helped to translate it to English.

References:
1. Kokryakov VN, Aleshina GM, Shamova OV, Orlov DS, Andreeva YuV (2010) Modern concept of antimicrobial peptides as molecular factors of the immunity. Med Acad Journ 10(4): 149-160. (in Russian)
2. Budikhina AS, Pinegin BV (2008) Defensins - multifunctional cations peptides of human. Immunopathology, Allergology, infectologist 2: S31-S40. (in Russian)
3. Zemko V, Okulich V (2013) Levels of enzyme activity of blood serum for the ability to destroy peptidoglycan in patients with purulent-inflammatory diseases. Regional European congress of biomedical laboratory science and the 4th Greek medical laboratory technologists conference. Athens.
4. Zemko V, Kiriluk O, Okulich V (2013) Diagnostics of purulent-inflammatory diseases by studying of enzyme activity of blood serum for the ability to destroy peptidoglycan. Actual problems of biochemistry and biotechnology 4: 171-173.
5. Zemko V, Kiriluk O, Okulich V, Senkovich S (2013) Method for isolating cell wall of Gram-negative bacteria: rats. Predlozhenie reg 31. (in Russian)
6. Zemko V., Kiriluk O., Okulich V. (2013) Act of introduction into the learning process ''Determing of enzyme activity of serum that destroy peptidoglycan'' 1 (in Russian)
7. Zemko V., Okulich V. (2013) Analysing of enzyme activity of blood serum for the ability to destroy peptidoglycan in patients with purulent-inflammatory diseases. Modern problems of infectious human pathology. Minsk. (in Russian)

Vitamin D: the foundation of Human Innate Immunity

A.S. Kapse

Professor, Department of Pediatrics, Mahavir super speciality hospital. Surat 395001 India

Medical scientists way back in 17th century had observed that nutritional rickets the prototypical disorder of vitamin D deficiency had close association with infections, indeed, many clinical scholars of that era had hypothesized that infection actually caused rickets. Robert Koch sought an infectious agent as the primary etiologic factor of rickets. Howland and Holt, major scholars of rickets, coined the term "rachitic lung" suggesting alliance of frequent pulmonary infections with vitamin D deficiency. Sir William Jenner described the association of rickets with TB. The high prevalence of TB, pneumonia and rickets was attributed to smog & crowding consequently it was seductive to consider infection the likeliest possibility.

By the mid 1920s, cod liver oil emerged as the effective treatment for rickets; therapeutic success was attributed to some "antirachatic factor" contained in fish oil. Effective element was later on identified as cholecalciferol & was termed as vitamin D. Manville suggested that fat-soluble vitamin deficiency "leads to frequently pyogenic infections of the respiratory tract and with this "the infection hypothesis" became less secure, and it was appreciated that it was vitamin D deficiency that led to infection.

The research in last decade has unearthed many new facets of vitamin D; the most important among them is its role as immunomodulator. The earlier observations that children with rickets are more likely to develop pneumonia or tuberculosis (TB) were augmented by new findings of increased propensity for respiratory viral infections with low vitamin D levels. Recent explorations establish that vitamin D and its metabolites operate at tissue, cellular, and nuclear sites and there by influence immune function at a subcellular level. This review is a humble attempt to depict emerging concepts of vitamin D as an immunemodulator.

Vitamin D the great Immune Modulator:

In the last decade scientists have sought mechanisms to account for the non-osseous effects of vitamin D. *Calcitriol (1, 25(OH) 2D) the active form of Vitamin D, is appreciated as a gene transcription factor.* Calcitriol (1, 25(OH) 2D), binds to a vitamin D receptor (VDR), this complex is then translocated to an intranuclear site where it regulates transcription of many proteins (>200). These proteins along with calcium and phosphate homeostasis also influence cell proliferation, cell differentiation, and importantly immune function.

All the cells of immune system (Dendritic cells, macrophages, and T and B cells) express VDR and CYP27b1 enzyme. This enzyme converts 25(OH) D into 1, 25 (OH) 2D the active vitamin D metabolite. This expression may be constitutive or induced post-stimulation. Expression of VDR and CYP27b1 increases manifold after immune stimulus. 1, 25(OH) 2D appears to act on all immune-related cells. *This active form of vitamin D influences immune cells genes and thereby regulates their various functions.*

Influence on Innate Immunity:

Evolutionarily innate immune system is one of the most crucial aspects of life protecting any organism from invaders. Recent studies have revealed potent effects of vitamin D on all the aspects of immune system, but most of the prominent advances in the past 5 years have focused on the ability of vitamin D to promote innate antimicrobial responses.

Innate Immunity: Barrier function.

The innate immune system begins with the epithelial barrier between the bacteria abundant in the outside environment and the effectively sterile host; epithelial cells in the skin, gut, respiratory and urinary tract, protect us from injury or invasion by infection. Vitamin D has a pivotal role in maintaining this physical barrier, the active hormone 1,25(OH)2D is important in upregulating genes via the CYP27b1 enzyme, which then encode proteins required for tight junctions (e.g. occludin), gap junctions (e.g. connexion) and adherens junctions (e.g. E-cadherin).

Innate Immunity: pathogen recognition.

Microbes have pathogen associated molecular patterns (PAMP's). These PAMP's are relatively invariant and are shared by many organisms however not by the host. When PAMP gain entry in to the host, they are recognized by a class of receptors in the plasma membrane of macrophages & epithelial cells known as Toll-like receptors (TLRs). There are around twelve different pathogen recognition receptor's (TLR) in mammals; they are located on many cells including macrophages, dendritic cells and epithelial cells. TLR2/1 binds to peptidoglycans found on Gram-positive bacteria such as streptococci and staphylococci. TLR4 is activated by LPS found on gram-negative bacteria, salmonella and *Escherichia coli*.

In humans, TLR2/1 and TLR4 when triggered result in the induction of the CYP27B1 (1a-hydroxylase) enzyme. This in turn induces the production of active vitamin D (1, 25(OH)2D). The 1,25(OH)2D binds to the vitamin D receptor which then bind to vitamin D-response elements unlocking the DNA, targeting genes that encode antimicrobial peptides.

Innate Immunity: Antimicrobial Peptides.

Human body produces human antimicrobial peptides in response to pathogens. These AMPs include: 6 human alpha-defensins (HNP1-4 and HD-5, 6), 4 human beta-defensins (hBD-1to 4), and cathelicidin (LL-37).

Vitamin D acts as a potent stimulator of antimicrobial peptides particularly cathelicidin and some defensins (defensins hBD-2) in the human body. Local production of 1, 25(OH)2D, the active vitamin D hormone in various tissues, is the preferred mode of response to antigenic microbial challenges. Calcitriol (1, 25(OH)2 D) has the ability to induce expression of cathelicidin in bronchial epithelial cells, urogenital epithelial cells, keratinocytes and myeloid cell lines. However macrophage & monocytes are the most important cells involved in cathelecidin creation. Local injury or infection in most epithelial sites results in liberation of cathelecidin.

Apart from its antimicrobial properties, cathelecidin also has other immune regulatory properties. Cathelicidin influences many other functions including chemotaxis, cytokine and chemokine production, cell proliferation, vascular permeability, angiogenesis, wound healing, etc. The response time for cathelecidin production is very short; within few minutes of pathogen recognition this protein is manufactured as defence mechanism.

Role of vitamin D in cathelecidin production:

Production of AMP particularly cathelecidin is dependent on sufficient circulating 25(OH)D.
In a series of studies carried out at the University of California, Los Angeles, Robert Modlin and his group showed that: (1) vitamin D sufficient individuals produced much higher levels of TLR1–TLR2-induced cathelicidin than a vitamin D insufficient person. (2) Addition of post vitamin D supplemented serum result into higher levels of cathelicidin production compared with serum obtained before treatment. (3) RNA interference (RNAi) knockdown of expression of cathelicidin abrogates the ability of 1, 25-dihydroxyvitamin D to promote intracellular microbial killing.

Cathelecidin production by Non Immune cells.

All immune cells produce cathelecidin but many of the non immune cells also produce this important AMP.
Epidermal keratinocytes:
Similar to monocytes epidermal keratinocytes are also capable of 1, 25-dihydroxyvitamin D-induced expression of cathelicidin, however in keratinocytes this process requires an initial trigger by transforming growth factor β1 (TGF-β1) to promote optimal expression of enzyme 25-hydroxyvitamin D-1α-hydroxylase and local synthesis of 1,25-dihydroxyvitamin D. Epidermal wounding is the classical example when the release of TGF-β1 initiates the above mentioned process. In addition to its antimicrobial activity, cathelicidin is known to modulate a wide range of immune responses, including effects on antigen presentation and epidermal wound-healing. These alternative functions of cathelicidin mean that its intracrine induction by vitamin D may have multiple beneficial consequences.
Placenta:
Vitamin D mediated intracrine induction of cathelicidin occurs in other human tissues, including the *decidual* (maternal) and *trophoblastic* (fetal) cells from the placenta.
Abundance of VDR in decidual and trophoblastic cells is consistent with localized responses to endogenous 1, 25-dihydroxyvitamin D. The placenta forms a natural barrier to fetal infection during pregnancy and both its maternal and fetal tissues are known to express antimicrobial proteins. A pivotal function of 1, 25-dihydroxyvitamin D production in the placenta might be to support the relatively high basal expression of antibacterial proteins to combat infection by pathogens such as *Listeria monocytogenes* and Group B Streptococcus that are known to have a role in adverse events associated with pregnancy.
Barrier sites:
Vitamin D induced antibacterial activity is probably a feature of cells at many so-called barrier sites, and this hypothesis has been endorsed by studies involving various cell lines. In particular, induction of cathelicidin by

1, 25-dihydroxyvitamin D in respiratory epithelial cells, has been linked to enhanced killing of airway pathogens such as *Bordetella bronchiseptica* and *Pseudomonas aeruginosa*. Similar responses in gastrointestinal epithelial cells at sites such as the gastrointestinal tract, cooperative stimuli may be required to initiate vitamin D induced bacterial killing. Moreover, as outlined earlier, vitamin D probably promotes tissue-specific, innate antibacterial mechanisms that are distinct from its effects on antimicrobial peptides.

From the recent vitamin D research cathelecidin has emerged as the most important tool of the human innate immunity which defends human body against varied microbes. Adams and his group have postulated that this primitive, non-endocrine biological system has evolved to control immune responsiveness to invading antigens. They opined that this primitive immune system is specific for primates and is acquired through acquisition of Alushort a genetic transponosome during evolution; moreover they claimed that phylogenetically this system pre-date the vitamin D endocrine system that controls divalent mineral balance.

Innate Immunity: Autophagy.

Autophagy is a eukaryotic mechanism that involves encapsulation of organelles or cell proteins in a double-membrane autophagosome prior to fusion with lysosomes. Subsequent degradation of the autolysosomal contents is a pivotal factor in normal cellular response to infection. Recent reports have observed that induction of autophagy is an essential feature of 1, 25-dihydroxyvitamin D induced human monocyte antimicrobial response to intracellular organisms like *M. tuberculosis* infection. Intracellular microbial killing is cathelecidin dependent; the effect is abrogated following RNAi knockdown of cathelicidin.

Innate Immunity: superoxide anions.

Besides its role in production of AMPs vitamin D also persuades hydrogen peroxide secretion in human monocytes. Activated by 1, 25(OH) 2D there is an increased oxidative burst which helps in killing the intracellular organisms.

Innate immunity: Cytokines regulation.

Vitamin D is essential in stimulating antigen specific T-cell activation should the innate immune system fails to control infection. Vitamin D primarily has a regulatory control on adaptive immunity so that it may prevent an overreaction of the inflammatory response in the adaptive immune system preventing further cell or tissue damage by inflammation. Vitamin D suppresses inflammation by limiting excessive production of proinflammatory cytokines like TNF-alpha and IL-12. Thus in an innate immune setting, vitamin D fulfils a dual function by promoting antimicrobial response to infection whilst helping to prevent an over elaboration of general inflammation.

Adaptive immune response:

Dendritic cells (DC) and macrophages, the two specialized cells in antigen presentation, initiate the adaptive immune response; they activate T and B lymphocytes. Activation & proliferation of T & B cells produce wide repertoire of immune responses. Importantly, the type of T cell activated, CD4 or CD8, or within the helper T cell class Th1, Th2, Th17, Treg, and subtle variations of those, is dependent on the context of the antigen presented by which cell and in what environment. *Among the many of factors which influence this process vitamin D is the most vital one.*

Conclusion:

Vitamin D plays the pivotal role for the proper functioning of the body's immune system; adequate levels of Vitamin D are essential for barrier integrity, chemotaxis, regulation of inflammation, & most importantly *"the production of antimicrobials peptides"*. Human immunity (both innate and adaptive immune system) entirely depends on sufficient vitamin D levels.

References for Lecture

[1] Wesley J. The Vitamin D Receptor: New Paradigms for the Regulation of Gene Expression by 1,25 Dihydroxyvitamin D3. Endocrinol Metab Clin N Am 39 (2010) 255–269

[2] Samuel S. Vitamin D's role in cell proliferation and differentiation; Nutrition Reviews® Vol. 66(Suppl. 2):S116–S124

[3] Ramgopalan SV: A ChIP-seq defined genome-wide map of vitamin D receptor binding: Associations with disease and evolution: http://www.genome.org/cgi/doi/10.1101/gr.107920.110

[4] Hewison M. Vitamin D and the Immune System: new perspectives on an old theme; Endocrinol Metab Clin N Am 39 (2010) 365–379

[5] Lorena Alvarez-Rodriguez: Age and low levels of circulating vitamin D are associated with impaired innate immune function; *J. Leukoc. Biol.* 91: 829–838; 2012.

[6] Hewison M. Vitamin D and the intracrinology of innate immunity; *Mol Cell Endocrinol.* 2010 June 10; 321(2): 103–111.

[7] Philips TL, Modlin RL. Toll-Like Receptor Triggering of a Vitamin D–Mediated Human Antimicrobial Response; www.sciencexpress.org / 23 February 2006 / Page 1/ 10.1126/science.1123933

[8] Lehrer, R. I., and Ganz, T. (2002) Cathelicidins: a family of endogenous antimicrobial peptides. Curr. Opin. Hematol. 9, 18–22

[9] Di Nardo, A., Vitiello, A., and Gallo, R. L. (2003) Cutting edge: mast cell antimicrobial activity is mediated by expression of cathelicidin antimicrobial peptide. *J. Immunol.* 170, 2274–2278

[10] Sorensen, O., Arnljots, K., Cowland, J. B., Bainton, D. F., and Borregaard, N. (1997) The human antibacterial cathelicidin, hCAP-18, is synthesized in myelocytes and metamyelocytes and localized to specific granules in neutrophils. *Blood* 90, 2796–2803

[11] Di Nardo, A., Vitiello, A., and Gallo, R. L. (2003) Cutting edge:
a. mast cell antimicrobial activity is mediated by expression of
b. cathelicidin antimicrobial peptide. *J. Immunol.* 170, 2274–2278

[12] Baroni E. VDR-dependent regulation of mast cell maturation mediated by 1,25-dihydroxyvitamin D3 *J.Leukoc. Biol.* 81: 250–262; 2007.

[13] Frohm, N. M., Sandstedt, B., Sorensen, O., Weber, G., Borregaard, N., and Stahle-Backdahl, M. (1999) The human cationic antimicrobial protein (hCAP18), a peptide antibiotic, is widely expressed in human squamous epithelia and colocalizes with interleukin-6. Infect. Immun. 67, 2561–2566

[14] Bals, R., Wang, X., Zasloff, M., and Wilson, J. M. (1998) The peptide antibiotic LL-37/hCAP-18 is expressed in epithelia of the human lung where it has broad antimicrobial activity at the airway surface. *Proc. Natl. Acad. Sci. USA* 95, 9541–9546

[15] Jürgen Schauber. Injury enhances TLR2 function and antimicrobial peptide expression through a vitamin D–dependent mechanism; J. Clin. Invest. 117:803–811 (2007).

[16] Gombart AF. Human cathelicidin antimicrobial peptide (CAMP) gene is a direct target of the vitamin D receptor and is strongly up-regulated in myeloid cells by 1,25- dihydroxyvitamin D3; *FASEB J.* 19, 1067–1077 (2005)

[17] Deretic, V., and Levine, B. (2009). Autophagy, immunity, and microbial adaptations. Cell Host Microbe 5, 527–549

[18] Geisler C. Vitamin D controls T cell antigen receptor signaling and activation of human T cells: 7 March 2010; doi:10.1038/ni.1851

[19] Oscar Palomares. Role of Treg in immune regulation of allergic diseases; Eur. J. Immunol. 2010. 40: 1232–1240.

[20] Prietel B. Vitamin D Supplementation and Regulatory T Cells in Apparently Healthy Subjects: Vitamin D Treatment for Autoimmune Diseases?; IMAJ 2010; 12: 136–139.

[21] Chen, S., Sims, G. P., Chen, X. X., Gu, Y. Y., Chen, S., and Lipsky, P. E. (2007). Modulatory effects of 1,25-dihydroxyvitamin d3 on human B cell differentiation. J. Immunol. 179, 1634–1647

[22] Hughes DA. Vitamin D and respiratory health. doi:10.1111/j.1365-2249.2009.04001.x

[23] Devereux G. Maternal vitamin D intake during pregnancy and early childhood wheezing; Am J Clin Nutr 2007;85:853–9.

[24] Camargo CA. Cord-Blood 25-Hydroxyvitamin D Levels and Risk of Respiratory Infection, Wheezing, and Asthma; doi:10.1542/peds.2010-0442

[25] Belderbos. Cord Blood Vitamin D Deficiency is Associated With Respiratory Syncytial Virus Bronchiolitis; DOI: 10.1542/peds.2010-3054

[26] Bergman P, Norlin A-C, Hansen S, et al. Vitamin D3 supplementation in patients with frequent respiratory tract infections: a randomised and double-blind intervention study BMJ Open 2012;2:e001663. doi:10.1136/bmjopen-2012-001663

[27] Hansdottir. Vitamin D Decreases Respiratory Syncytial Virus Induction of NF- kB-Linked Chemokines and Cytokines in Airway Epithelium While Maintaining the Antiviral State; *J Immunol* 2010;184;965-974

[28] Xystrakis E. Reversing the defective induction of IL-10–secreting regulatory T cells in glucocorticoid-resistant asthma patients; J. Clin. Invest. 116:146–155 (2006)

[29] Philips TL, Modlin RL. Cutting Edge: Vitamin D-Mediated Human Antimicrobial Activity against Mycobacterium tuberculosis Is Dependent on the Induction of Cathelicidin

[30] Martineau, A. R A single dose of vitamin d enhances immunity to mycobacteria. Am. J. Respir. Crit. Care Med. 176, 208–213

[31] Gal-Tanamy M. Vitamin-D: An innate antiviral agent suppressing Hepatitis C virus in human hepatocytes; Hepatology 2011 Jul 25. doi: 10.1002/hep.24575.

[32] Mehta S. Low Maternal Vitamin D Increases Risk of HIV Transmission to Offspring; J Infect Dis 2009;200:1022-1030

[33] Emilio Sánchez-Valdéz. Clinical Response in Patients With Dengue Fever to Oral Calcium Plus Vitamin D Administration: Study of 5 Cases

[34] Hossein-nezhad A, Spira A, Holick MF (2013) Influence of Vitamin D Status and Vitamin D3 Supplementation on Genome Wide Expression of White Blood Cells: A Randomized Double-Blind Clinical Trial. PLoS ONE 8(3): e58725. doi:10.1371/journal.pone.0058725